BRICKWORK & BRICKLAYING

A DIY HANDBOOK

BRICKWORK & BRICKLAYING

A DIY HANDBOOK

Jon Collinson

THE CROWOOD PRESS

First published in 2012 by
The Crowood Press Ltd
Ramsbury, Marlborough
Wiltshire SN8 2HR

enquiries@crowood.com
www.crowood.com

Paperback edition 2024

British Library Cataloguing-in-Publication Data
A catalogue record for this book is available from the British Library.

ISBN 978 0 7198 4468 3

All illustrations by the author unless otherwise stated.
Frontispiece courtesy of Hanson Building Products.

Cover design by Nautilus Design (UK) Ltd
Typeset by Jean Cussons Typesetting, Diss, Norfolk
Printed and bound in India by Parksons Graphics Pvt Ltd

Contents

Preface

I began as a bricklayer in 1984, on the recommendation of my father (one rainy Tuesday afternoon, as I recall, in Nottingham's Victoria Shopping Centre), who saw the practical and financial worth of such a course of study, and my career path has stayed within the construction industry ever since.

It has been a varied career, which has encompassed bricklaying, health and safety, estates management and many years of teaching brickwork as a vocational craft subject. In all those years, however, I have never found one craft book, among the many excellent works by expert practitioners, which gives someone new to bricklaying all the underpinning and practical knowledge needed in one simple, accessible volume, at the same time providing a basis for further reading. It is with this in mind that I decided to write this book! It is intended to provide an overall appreciation of the materials and the basic practical skills associated with the craft, to enable the reader to undertake simple practical bricklaying projects of his or her own.

Acknowledgements

I would like to acknowledge the fundamental part played by the following people in shaping my knowledge, experience and understanding over the years, both as a student and beyond. Without them this book could not have been written. In addition, the list also gratefully acknowledges those individuals who have provided practical support and assistance with both information and images used in the book: Jim Beardsley, Hugh Carr, Ken Crowe, Albert Johnson, Ray Linley and Andy Williamson (all South East Derbyshire College); Steve Morton (Hanson Building Products); Robert Parkin (North Nottinghamshire College); Andy Thomas (Travis Perkins Eastwood); Annabelle Wilson (Ancon Building Products); and Nicola Jane Slack, for everything!

Thanks are also given to the numerous, unknown bricklayers whose work is included in this book in order to illustrate good practice.

Finally, this book is dedicated to the memory of my late father, John Foster Collinson, who set me on the path for a most rewarding career and who was a 'gentleman builder' of the old school for whom 'It'll do' never would!

CHAPTER 1

Introduction

The use of bricks and the ancient craft of bricklaying have been in existence for thousands of years. The oldest shaped mud bricks, discovered near Damascus in Syria, date from as far back as 7500BC. Mud bricks were extensively used by the civilization of Ancient Egypt and the first sun-dried clay bricks date back to 4000BC, having been discovered in Mesopotamia (now Iraq). The Chinese also were experts in stonemasonry and bricklaying, the most iconic example of their work being the Great Wall of China, which was begun in the fifth century BC and is claimed to be the only man-made object visible from outer space.

The Romans made use of fired bricks and the Roman legions, who were known to operate mobile kilns, introduced bricks to many parts of their empire, including Europe around 2000 years ago. Great innovators in many areas, the Romans developed bricklaying as a craft, including the use of mortar and different types of bonding arrangements; however, with the eventual decline of the Roman Empire, the craft of bricklaying declined with it.

It was not until the latter half of the seventeenth century, after the Great Fire of London in 1666 in fact, that the English again started to use bricks in building and it took almost another 200 years, until the middle of the nineteenth century, before the mechanized production of bricks began to replace manual methods of manufacture. Despite the advent of mechanized production, however, growth in the brick industry was relatively slow as the moulded clay bricks were still being fired in fairly inefficient static or intermittent kilns. In 1858, a kiln was introduced that allowed all processes associated with firing the bricks to be carried out at the same time, and continuously. Since the introduction of this, the Hoffmann kiln, the brick industry has made great progress, particularly since 1930, when the output of bricks in Great Britain doubled up to the start of the Second World War.

Clay has provided the basic material of construction for centuries and brick properties vary according to the purpose for which they are intended to be used. Today, clay bricks feature in a wide range of buildings and structures, from houses to factories. They are also used in the construction of tunnels, waterways and bridges, and so on. Many hundreds of attractive varieties, colours and texture of brick are available, which can be used imaginatively and creatively to greatly enhance the physical appearance and design of modern buildings.

CHAPTER 2

Concrete

Concrete is a mixture of cement (usually Ordinary Portland cement), acting as a binder, fine aggregates such as sharp sand, and coarse aggregates such as gravel or crushed stone. Alternatively, fine and coarse aggregates may come pre-mixed as all-in ballast. Water is added to form a paste with the cement, which covers the surface area of every stone and aggregate particle in the mix, binding them together to form a solid mass when the concrete hardens. The aggregates are not physically altered in any way, but are firmly set into the hardened, rock-like cement paste.

Concrete has numerous applications, among them the foundations of buildings, oversite concrete under suspended timber ground floors, solid floor slabs, driveways, paths, benching in man-holes, and many others.

In some cases a chemical plasticizer is mixed with the water in order to make the concrete easier to work and to protect it from frost while it sets.

CEMENT

Cement is generally regarded as the most important binding material used in the construction process and is used for the manufacture of mortar and concrete. The use of cement is so common that it is the second most consumed substance in the world after water! The most commonly used

> **Hydration**
>
> The reaction between the cement and water is an exothermic (meaning 'giving off heat') chemical reaction known as 'hydration', which results in the setting of the cement paste. Concrete, and mortars for that matter, set as a result of the completion of the hydration process and not simply by drying out.

Fig. 1 Ordinary Portland cement.

Fig. 2 A 25-kg bag of Ordinary Portland cement.

Cement Hazards

Cement is classified as an irritant, having the potential to cause severe cases of dermatitis and burning of the skin. Great care must therefore be taken when handling cement (and the materials made from it) in order to avoid contact with the skin or eyes and breathing in cement dust. The use of PPE (personal protective equipment) is essential.

is Ordinary Portland cement (OPC), so called because, in its solid state, its grey colour is much like that of natural Portland stone. The fact that Portland cements form a solid when mixed with water means they are often referred to as 'hydraulic cements'.

Cement is made from chalk (calcium carbonate) excavated from natural chalk deposits, which is then heated in a rotating cement kiln at temperatures of up to 1450 degrees centigrade. The resultant clinker is then ground to a fine powder, which is then packaged in 25-kg bags.

Although Ordinary Portland cement is the most commonly used for concrete and general construction purposes, there are various other types with slightly altered chemical compositions for specialized circumstances. These include rapid-hardening cement, sulphate-resisting cement, special cements for working in cold weather, and so on.

Being hydraulic, bags of cement must be protected from damp before use. Preferably, bags should be stored clear of the ground on a wooden pallet in a well-ventilated, rain-proof shed. Bags of cement should be stacked flat, no more than five bags high, otherwise the bags at the bottom will set due to the pressure exerted by the bags above. 'Pressure setting', as it is called, is sometimes known as 'warehouse setting'.

Even when stored in ideal conditions, cement still has a 'shelf life' and will lose around 20 per cent of its strength over a period of a couple of months. Accordingly, care must be taken to ensure that cement is used in the same sequence in which it was delivered – in other words, old bags first! Under no circumstances should cement that has been exposed to moisture and contains lumps be used as it will produce a weak and less durable mix.

FINE AGGREGATE (CONCRETING SAND OR SHARP SAND)

'Concreting sand', 'sharp sand' or 'fine aggregate' are all terms that are used to describe natural sand, crushed stone, sand or similar, which pass through a 5mm sieve but tend to be coarser and not as well graded as the soft sands used in mortars. It is often referred to as 'fine aggregate' to distinguish it from soft sand. Being coarser than soft sand, it is not used for mortars as it produces an unworkable mix and causes difficulties in achieving a good finish when jointing.

Fig. 3 Sharp sand.

COARSE AGGREGATES

Coarse aggregates are excavated from natural deposits of gravel or stones and then washed to remove dirt, salts and clay. The term describes materials such as natural gravel, crushed gravel or crushed stone that will not pass through a 5mm sieve. Bigger stones are often crushed to make

Fig. 4 Coarse aggregate.

smaller pieces. Washing, crushing and sieving usually take place in the location of the excavations. A coarse aggregate used for concrete is likely to pass through a 19mm sieve but stay on a 10mm sieve.

ALL-IN BALLAST

If the coarse aggregate and concreting sand/fine aggregate are purchased already mixed together, the material is known as 'all-in aggregate' or 'all-in ballast'. Aggregates for concrete can be purchased in 25-kg bags, 1-tonne 'dumpy bags' or loose by the lorryload for very large deliveries.

STORAGE OF AGGREGATES

One of the key factors when storing aggregates for concrete (or sand for mortars) is that they must be kept clean and dry. Water contained in aggregates and sands has implications for the water content of the finished mix, which in turn affects the workability and performance of the finished concrete or mortar. Contaminants such as soil, soluble salts from the ground, dirt, leaf matter, and so on, can cause staining or interfere with the bond between the cement paste and the aggregates in the mix.

Aggregates and sands delivered in 25-kg bags should be stored clear of the ground on a pallet. Inevitably, some bags will have been torn or punctured during original storage, transport or delivery, so it is good practice to cover the bags over with a tarpaulin or polythene. Deliveries made in 1-tonne 'dumpy bags' should, ideally, be kept off the ground on a pallet but if this is not possible they should be stored on a concrete or similar hard base. The most important factor is that they should never be in contact with soil. In all cases, the bags should be covered over with a tarpaulin or polythene to prevent rain ingress (or contamination by falling leaves during autumn).

Where loose deliveries are made by the lorryload, aggregates and sands should be stored on

Fig. 5 'All-in ballast'.

Fig. 6 A 25-kg bag of aggregate.

Fig. 7 One-tonne 'dumpy bag' of aggregate.

site on a hard, well-drained, sloping base. Ideally, the base should be made of concrete with the surface laid to a fall or slope to allow any water to drain out. On larger projects, purpose-built blockwork bays are constructed to provide separate storage for different aggregates. Again, the bays should be covered over with a tarpaulin or polythene.

GRADING OF AGGREGATES

'Grading' refers to the size distribution of particles within a batch or sample of aggregate. The particle size (and hence the surface area) of an aggregate is a very important factor in designing the mix proportions for concrete. In order for concrete to attain its maximum strength, all surfaces of the aggregates must be coated with cement paste. Since the surface area of one grain of sand is less than that of two grains of half the size, it follows that a concrete mix should not contain too many fine particles. Particles that are too fine will contribute to a weaker mix due to the increased surface area needing to be coated with cement paste. Alternatively, an aggregate such as uncrushed stone that is more uniform, containing particles of similar size and lacking finer particles, will require more cement paste to fill the voids that would have been filled by fine particles found in better-graded aggregates. The use of uniformly graded aggregates will result in a mix that lacks workability, and weak, porous finished concrete.

Clearly, when well-graded aggregates are used, containing the full range of particle sizes, with the smaller aggregates filling the voids between the larger stones, a strong, workable concrete mix can be produced with the minimum of cement.

AGGREGATE/CEMENT RATIO OF CONCRETE MIXES

There are two key considerations when designing concrete mixes for different applications: the ratio of aggregate to cement and the ratio of water to cement.

The aggregate/cement ratio relates to the proportions of dry materials in the mix. The characteristics of finished concrete, including its strength, can be directly affected by the amount of cement in relation to aggregates. Mix proportions are usually specified as a ratio of cement to fine aggregate to coarse aggregate, either by weight or by volume. For example, a 1:3:6 mix contains one unit of cement, three units of fine aggregate and six units of coarse aggregate. It is vital that there is sufficient cement within a mix to bind all of the aggregates together, otherwise weak patches will occur in the finished concrete. On this basis, the higher the cement content in relation to aggregates, the stronger the concrete. For example, a mix of 1:1:2, proportionally having a very high cement content, will produce very strong, impermeable, frost-resistant and durable concrete.

As mentioned previously, fine aggregate and coarse aggregate can be purchased together as all-in ballast but it should not be assumed, when proportioning by volume, that 4m^3 of fine aggregate when mixed with 8m^3 of coarse aggregate will produce 12m^3 of all-in ballast. Volume shrinkage upon mixing of around 30 per cent will occur, caused by the finer particles filling the voids within the coarse aggregate. The resulting volume of all-in ballast will be around 9m^3. Accordingly, great care must be taken when specifying the quantity of aggregates.

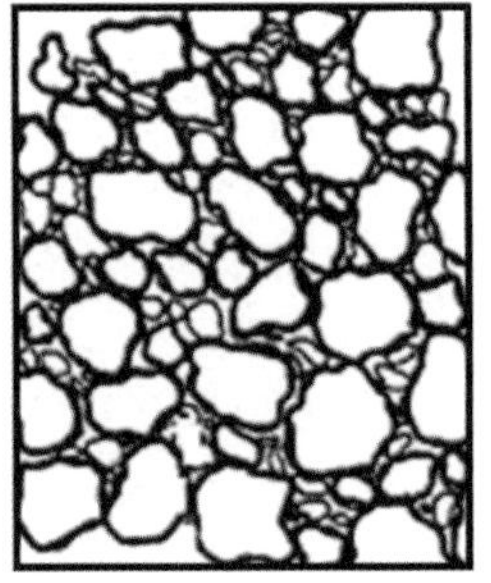

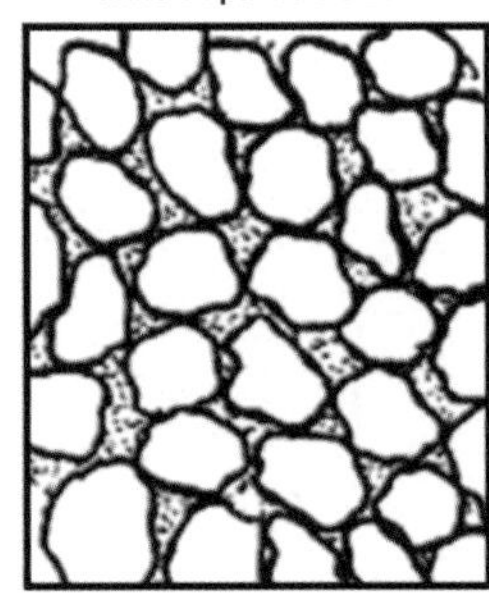

Fig. 8 Grading of aggregates.

Typical Concrete Mix Proportions

Cement	*Fine agg'*	*Coarse agg'*	*Application*	*Cement*	*All-in ballast*
1	½	1	Benching in a manhole	1	1
1	1	2	Chimney flaunching	1	2
1	2	4	General exterior work, casting in moulds, concrete patching	1	3 or 4
1	3	6	Floor slabs and foundations	1	5 or 6

GAUGING OF CONCRETE MIX MATERIALS

In this context, 'gauging' is the term used to describe the measuring out of quantities of individual dry materials for a concrete mix. It is a common practice – and a bad one – on many construction projects simply to load a concrete mixer using a shovel, with the result that there is no accurate control in terms of gauging the proportions of materials. For example, one shovelful of damp sharp sand has greater volume than one shovelful of dry cement powder, so mixes batched in this way often contain too little cement. Realistically, the only way that a degree of accuracy can be employed is to use batching by weight or by volume. The same principles apply to measuring out materials for mortar mixes.

Weight Batching

Weight batching is the more accurate of the two methods used for gauging the proportions of materials. It is more likely to be found only on large construction sites as it involves sizeable mixing machines fitted with loading hoppers, or separate weight-batching hoppers that discharge directly into a large concrete mixer.

Volume Batching

Volume batching is the most suitable method used on small projects or for smaller quantities of concrete where mixing might be carried out by hand or by using a manually loaded concrete mixer.

The accuracy of this method relies on the use of a standard unit of volume for gauging the quantity of each material. A simple example is a bucket that can be filled to the top with each material and emptied into the mixer the relevant number of times as determined by the mix proportions. A quicker method is to use a 'gauge box', designed to the required size. It is a bottomless box into which the individual dry materials are placed before being ruled off with a timber straight-edge at the top in order to get equal proportions.

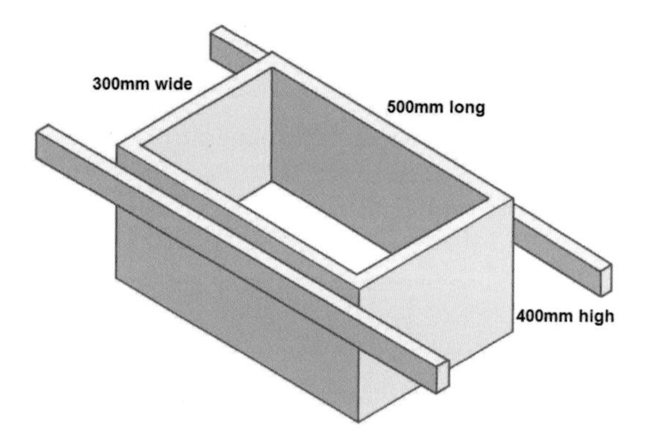

Fig. 9 A typical example of a gauge box.

The dimensions of the box in Fig 9 equate to four 25-kg bags of cement but smaller boxes for smaller projects are available, or could easily be made out of plywood for smaller mix requirements.

The gauge box is placed on a clean, firm base (wooden board or steel sheet) and filled in proportion to the design of the mix. When full, the box is lifted and removed, and the contents can be shovelled into a mechanical mixer or on to an appropriate hard surface for mixing by hand.

WATER/CEMENT RATIO OF CONCRETE MIXES

The first essential criterion when mixing concrete is to ensure that the water is clean, as any impurities will affect the strength of the finished concrete.

> **Volume Shrinkage**
>
> When volume batching for concrete, account must be made for 'volume shrinkage' on mixing. For example, $1m^3$ of cement, $3m^3$ of fine aggregate and $6m^3$ of coarse aggregate will not produce $10m^3$ of finished concrete. Because the small particles will fill the voids between the bigger particles, the actual volume will be around $7m^3$. Volume shrinkage of concrete materials is around 30 per cent upon mixing, so care must be taken when specifying quantities for concrete!

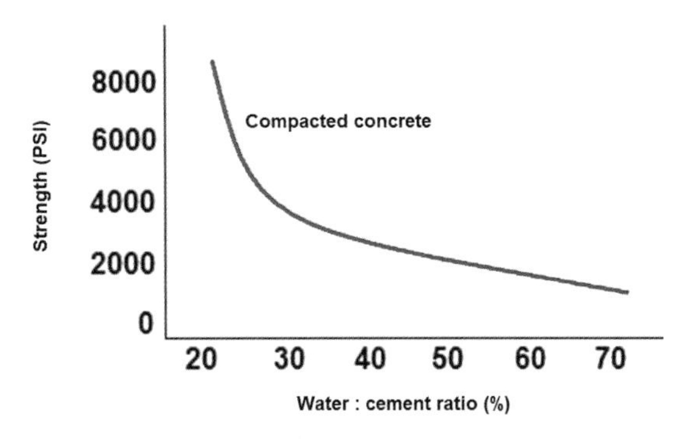

Fig. 10 Concrete strength related to water:cement ratio; the strength of finished concrete decreases as the water content of the mix is increased.

The term 'drinkable water' is often used to define the quality of clean and uncontaminated water that is fit for use in concrete. Water fit for drinking is also referred to as being 'potable'.

The quantity of water used in mixing concrete is a very important consideration as the amount of water in the mix has a direct bearing on the properties of the finished concrete.

The real job of water is to make the cement set hard – to change it from a powder to a solid mass (in the process of hydration), which binds all the aggregates together. The process actually uses only a small part of the water added to concrete although insufficient water will not chemically 'activate' the cement and/or will reduce the ability to compact the concrete properly. The rest of the water is there to make the concrete workable enough to be transported, placed and compacted properly. All the extra water will evaporate out as the concrete sets and each drop will leave behind a tiny air pocket in its place, which renders the concrete a little like a solidified sponge. A higher volume of excess water leads to more holes, and a weaker concrete that will be more susceptible to frost damage. In any severely cold weather that follows, water inside those air pockets will expand on freezing and cause the concrete to crack and/or the surface to spall.

Excessive water results in excessive workability and elongated setting times but can also dilute the cement paste to the point where it drains away from the aggregate, causing the whole mix to segregate. Also, the more excess water present, the greater the tendency for it to rise to the surface of newly placed concrete. This 'bleeding' forms fine, open channels, which remain after the concrete has set, and reduce its durability and frost resistance.

On average, the water content should be 50 per cent of the weight of the cement in the mix to give the best balance between workability and strength. This is referred to as the water/cement ratio. It must be remembered that, when adding water to the concrete mix, an allowance must be made for any water that might already be present in the aggregates. This is why it is always best to store aggregates in such a way that they stay dry.

MIXING CONCRETE

Mixing Concrete by Hand

Small quantities of concrete will often be mixed by hand. The area selected for mixing must be flat, hard, and free from debris and dust. A concrete

> **Water Content**
>
> The water content of concrete is a vitally important factor: too little and the concrete is hard to compact and the hydration process may not take place satisfactorily; too much and workability becomes excessive, setting time increases and the durability, strength and frost resistance of the finished concrete diminish.

floor or patio area is ideal but the following factors must also be considered:

- Concrete will inevitably stain the mixing surface so an area should be chosen where this does not matter. Staining will still occur even if the area is hosed down immediately afterwards.
- The area should be large enough to use the mixing shovel and turn the materials easily, with sufficient space to move around. Preferably, the area should be close to the place where mixing materials are stored and/or where the mixed concrete is required to be placed.
- It is important not to mix for too long as this can cause the mix components to segregate. All batches should be mixed for the same amount of time. Sufficient mixing time should be allowed to ensure that all aggregates are coated with cement paste.

Bespoke large plastic mixing trays with raised edges are available from builders' merchants but may be problematic to store later or an unnecessary expense if they are only to be used once or infrequently.

The following tools and equipment will be required for hand-mixing concrete:

- a wheelbarrow for transporting materials;
- a water butt, such as a plastic or metal drum or barrel, or alternatively access to an outside tap or hose;
- builders' plastic buckets, preferably three: one for water, one for cement and one for aggregates, since the latter are often damp. An alternative to a bucket is a gauge box;
- a builder's shovel for mixing.

Measure out the correct proportions of materials and place the dry ingredients together on the mixing area. Using a shovel, turn the mix over into a single pile, ensuring good integration of the dry materials. Repeat the process by turning the mix back again. The main objective of this is an even distribution of the cement within the particles of aggregate. Repeat this process a third time, by which point the dry mix should be fully integrated.

Make a hole in the centre of the dry mix, ready to receive the water from a bucket. It is vital to ensure that the water does not escape, taking cement with it and thereby reducing the strength of the mix. The materials should be mixed with the water from the outside into the middle until all the water has gone. Turn all the materials over on to the top of the heap and continue turning the whole mix over until it is fully mixed and workable. Use a shovel to place the mixed concrete into a barrow. Make sure the inside of the barrow has been wet first to assist with tipping out the mix later.

Mixing Concrete by Machine

There are many different types of mixers, from the portable mixers suitable for the small builder up to the very large static mixers suitable for very large sites where the concrete is being site mixed. Mixers can be powered by petrol or diesel, and the smaller portable ones by electricity.

Whenever a mixer is being used, all safety measures must be taken and the manufacturer's instructions followed:

1. Set the mixer up according to the manufacturer's instructions. It is always a good idea to use polythene to sheet down under and around the mixer, in order to minimize mess.
2. Ensure you have sufficient materials and small tools.
3. Start the mixer.

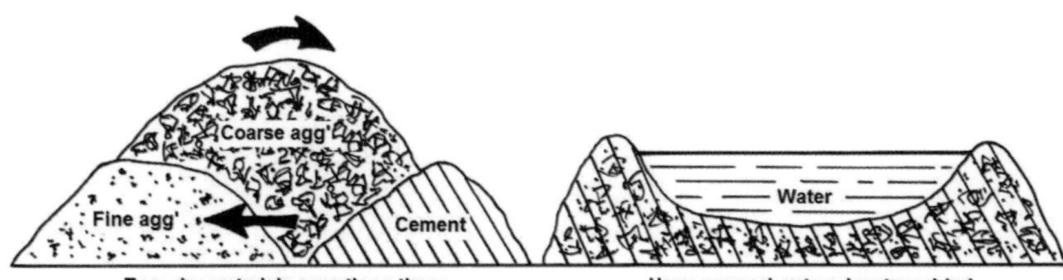

Fig. 11 Mixing concrete by hand.

Fig. 12 Electrically powered concrete mixer.

4. Add around half the water, using a bucket.
5. Add the pre-gauged materials or separate materials in the following order: half of the aggregate, then cement, then the other half of the aggregate.
6. Add more water as necessary, to achieve the required workability.
7. Allow the materials to mix for 3 minutes – any longer may cause segregation of the mix. Timing should be taken from when all of the materials are in the barrel of the mixer. All batches should be mixed for the same length of time.
8. Turn out the mix into a wheelbarrow – wetting the inside of the barrow will assist with tipping out the mix later.

READY-MIX CONCRETE (RMC)

Ready-mix concrete is manufactured in a factory or batching plant to specified mix proportions and then delivered to site by lorry-mounted tran-

> **Cleaning Mixing Areas**
>
> On completion of mixing operations, the mixing area should be thoroughly cleaned and tidied. If tools and equipment are not cleaned thoroughly after use, cement paste will harden on them and will be difficult to remove later. This is particularly important if pigments have been used in the mix.

sit mixers. This results in precise mix proportions being delivered in pre-ordered quantities (measured by volume in m^3), which can be custom-made to suit many different construction applications. Ready-mix concrete is manufactured under computer-controlled operations and transported and placed at site using sophisticated equipment and methods. Ready-mix concrete provides customers with numerous benefits but a few disadvantages.

The advantages of ready-mix over site-mixed concrete are as follows:

- a better and more consistent quality concrete is produced;
- there is no need for storage space for basic materials on site;
- the elimination of procurement/hiring of mixing plant and machinery;
- wastage of basic materials is avoided;
- the labour associated with site production of concrete is eliminated;

Fig. 13 Ready-mix concrete lorries.

- the time required is greatly reduced;
- noise and dust pollution on site are reduced.

However, ready-mix concrete does have a number of disadvantages:

- The materials are batched and mixed at a central plant, so the travelling time from the plant to the site is critical over longer distances. Some sites are just too far away!
- Concrete's limited lifespan between mixing and 'going off' means that ready-mix should be placed within 2 hours of batching at the plant. Concrete is still usable after this point but may not conform to relevant specifications.
- It generates additional road traffic. Furthermore, access roads and site access have to be able to accommodate the weight and size of the truck and load.
- Any shortfall in quantity, due to a miscalculation, could result in work already done being wasted. Conversely, over-ordering will result in costly wastage of unused concrete.

MINI-MIX CONCRETE

A smaller-scale alternative to ready-mix concrete is the use of contractors who provide 'mini-mix' concrete, arriving at site with lorries containing all the separate concrete constituents, including water. The lorry has a separate mixer with integrated weight-batching facilities, allowing precise, small quantities, up to approximately 6m^3, to be mixed on site to specified mix proportions. Mini-mix suppliers will also barrow the concrete to where it is required, within reason. Such a service eliminates many of the disadvantages associated with ready-mix concrete.

Fig. 14 Lorry delivering 'mini-mix' concrete.

PLACING AND COMPACTING CONCRETE

Ensuring the full and proper compaction of concrete when it is placed is of fundamental importance because any voids or pockets left in the finished concrete will compromise its final strength. As an indication, voids that total 5 per cent of the concrete by volume will reduce the strength of the concrete by as much as 30 per cent.

Concrete with minimal water will be high in strength but will need mechanical compaction, whereas a high water content will give workable concrete that is easily compacted by hand, but the concrete will be weaker. Very wet mixes are very easily compacted since the concrete will practically flow into position but, again, final strength and durability will be further compromised by the high ratio of water to cement.

Freshly mixed concrete goes stiff after 30 minutes. This is referred to as the 'initial set' and once concrete has reached this stage it should not be mixed up again. Beyond initial set the concrete hardens and rapidly gains strength. After about 10 hours the concrete is hard, having reached what is called the 'final set'. Having achieved its final set, concrete then slowly continues to harden and gain strength over many years and decades.

Accordingly, placing and compacting need to be done quite quickly, before the initial set takes place, especially where the concrete mix has a proportionally high cement content. Such a concrete will set more quickly than a 'lean mix' containing less cement.

For small concreting jobs, a short length of 50 × 50mm timber can be forcefully poked or rammed into the freshly placed concrete to remove any air pockets and ensure compaction.

For slabs and/or larger areas constructed within a frame of timber formwork, a heavy, straight timber plank (of up to 125mm × 50mm sectional size), preferably with raised handles or cut-out handles at the ends, can be operated by a man at

Fig. 15 Use of a tamping board to compact and level concrete within timber framework.

each side of the formwork. The 'tamping board', as it is known, is raised and lowered whilst gradually working all the way along the concrete surface to achieve compaction.

Where necessary, the tamping board can be 'shuffled' from side to side as it is drawn forward along the top of the formwork in order to grade off any excess wet concrete and to obtain a flat surface. Generally, the timber straight-edge used as a tamping board must be long enough to extend 100mm beyond both sides of the formwork in order to be effective. Hammering the exposed sides of the timber formwork will also assist compaction of the concrete at its perimeter sides.

The construction of formwork and shuttering is covered in greater detail in Chapter 5.

SURFACE FINISHING OF CONCRETE

In some situations, where a non-slip surface is required, the finish provided by the tamping process is left as it is and no more surface finishing is carried out. Where a smooth surface is required, finishing of concrete should take place straight away after placing and compaction. The surface should first be lightly trowelled over with a steel floating trowel. If this is done excessively a layer of laitance (very watery cement slurry) will be brought to the surface. This laitance makes subsequent floating of the surface difficult and leaves patches on the surface of the finished concrete that are porous and less durable.

Depending on the setting rate, the surface should be floated after 2–3 hours, using a wooden or polyurethane float in a circular motion. This will flatten any lumps or high spots and fill any dips or hollows in the surface whilst at the same time bringing some moisture to the surface.

At this point, a stainless-steel floating trowel can be run over the surface to provide a final smooth finish. Care should be taken not to over-trowel the finished concrete as this can bring more water to the surface, which will bring with it very fine particles of aggregate. When the concrete hardens, the fine, unbound particles are left behind on the surface, forming a dust when dislodged by any surface friction.

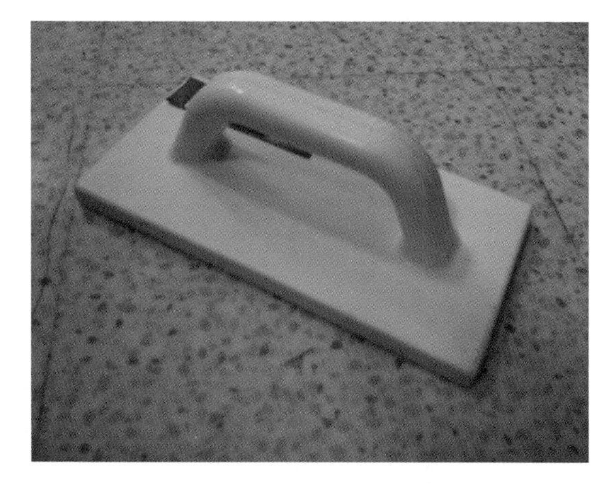

Fig. 16 Polyurethane float.

Fig. 17 Stainless steel floating trowel.

CURING AND PROTECTING NEW CONCRETE

Freshly placed and finished concrete needs to be kept moist. As water is essential to the hydration process taking place within the concrete, keeping it moist helps it retain, or if necessary absorb, the additional moisture it needs to complete the hydration, setting and hardening process. This process, known as 'curing', needs to be carried out carefully and properly if concrete is to maximize its hardness and strength.

Although it will achieve its final set in around 10 hours, the concrete is still comparatively weak at this stage. It will gain strength in the days that follow and achieve around 90 per cent of its final strength after approximately three weeks; it may then continue to strengthen for decades.

If concrete loses its moisture content too quickly, tensile stresses build up in it. The concrete is still too weak at this point to resist these stresses and both internal and surface cracks can result. These not only spoil the surface finish but also reduce the concrete's final strength, making it less durable and more susceptible to frost damage.

New concrete must also be protected from the sun and the wind, which will both speed up the evaporation process, causing shrinkage cracking, particularly at the surface. These problems are especially significant where concrete is being laid in thin sections. Mixing and placing concrete in such weather conditions will have the same injurious effect on the water content during these processes and should be avoided if possible.

It is vital to cure concrete properly in the first three days to increase its early strength development and long-term strength, impermeability and durability. Ideally, it should then be kept in conditions of controlled temperature and humidity. In practical terms, this is achieved by a number of methods, depending on the situation:

- Periodically spraying the concrete surface with water.
- Where possible, flooding or 'ponding' the entire surface – sometimes it may be necessary to build earth or clay dams around the area being cured.
- Covering the surface with hessian or coconut matting that is maintained in a wet state.
- Covering the surface with wet sand.
- Covering the surface with waterproof paper or polythene to retain the moisture.

Curing should begin as soon as the concrete surface is sufficiently hard that it will not be marked or damaged by the water spray or covering.

DRYING SHRINKAGE OF CONCRETE

When finished concrete hardens and effectively has dried out for the first time, its linear dimensions will contract by approximately 0.4mm per metre. This reduction in volume is primarily due to the excess water in the concrete (that used for workability as opposed to hydration) being yielded up to the atmosphere by way of evaporation. On subsequent wetting from rain, for example, it will expand again but it will not quite achieve its original dimensions.

The issue of drying shrinkage becomes more significant on larger areas of concrete. It can be minimized by reducing the amount of excess water in the original mix, provided that full compaction of the concrete can still be achieved. Whilst a reduced water/cement ratio may mean that the concrete is more difficult to place and compact, there is the advantage that the finished concrete will be stronger and less porous, which will decrease moisture movement in terms of swelling and contracting in wet weather.

The more quickly concrete is allowed to release its excess moisture content, the more it will shrink, so drying shrinkage can also be reduced by way of effective curing. This will allow the concrete to retain its moisture content whilst it gains strength for the first few days after placing.

WEATHER CONDITIONS

Concreting in Hot Weather

Many of the problems associated with concreting in hot weather are overcome by placing concrete quickly and preventing premature moisture loss by way of adequate curing. In addition, batches of mixed concrete should not be left standing during periods of hot weather so the timing of mixing and placing is important. Concrete generally, but particularly in hot weather, must be placed in a continuous, smooth operation. If one batch of concrete is left to stand and stiffen before the next batch is placed on or next to it, 'cold joints' will form and the concrete will not be one homogenous mass.

Aggregates should be kept cool by spraying with water but allowance must be made when mixing for the increased water content of the aggregate. In addition, or as an alternative, a chemical retarding agent can be added to slow the hydration process.

Ready-mix concrete should be kept agitating in hot weather but care must be taken not to overmix in case segregation occurs.

Curing should start at the earliest opportunity to avoid premature drying-out of the concrete surface and resultant surface cracking.

Concreting In Cold Weather

At low temperatures, the hydration process, setting and gaining strength within concrete are delayed, so protecting concrete from the cold for extended time periods is necessary. Curing periods are generally doubled during winter months but, even if the concrete can be kept above freezing temperatures, the strength of concrete laid at low temperatures can be as much as 25 per cent lower than that of concrete placed at warmer times of the year.

One of the key problems is that water expands when it freezes, which causes massive problems if freezing occurs within partly set concrete. At best, the concrete strength can be significantly reduced as a result of the freezing water inducing stresses that weaken the bond between the aggregate and the cement paste. This can also result in increased porosity, spalling of the concrete and loss of durability. At worst, the damage can be so severe that the concrete is rendered useless and must be taken up and replaced.

Possible courses of action include raising the temperature of the mixing water, ensuring that all aggregates are free from ice and not frost-bound, and protecting the finished concrete with polythene sheeting with hessian quilting on top, held down firmly at the edges. However, the best course of action is not to carry out concreting at temperatures below 5 degrees centigrade or where there is a risk of the temperature dropping below 5 degrees centigrade overnight.

The use of Rapid-Hardening Portland cement instead of Ordinary Portland has benefits for working in cold weather. Rapid-Hardening Portland cement and Ordinary Portland cement are similar in composition but the former is more finely ground. The increased surface area of the cement powder does not make concrete set more quickly, but it does increase the rate of hydration at early stages and leads to an increased rate of early hardening and strength gain. This is important when trying to combat the effects of cold weather on new concrete. The increased rate of hydration, being an exothermic reaction, also causes an increase in the evolution of heat, which helps to reduce the possibility of water freezing within the concrete.

CONCRETE CHEMICAL ADMIXTURES

Chemical admixtures are materials in the form of powder or fluids that are added to plain concrete mixes at the time of batching or mixing in order to give the finished concrete certain properties. In normal use, admixture content is usually less than 5 per cent by mass of cement. There are a number of common types of admixtures:

- Accelerators speed up the hydration and early hardening and strength gain of concrete, which is important when trying to combat the effects of concreting in cold weather.
- Retarders slow the hydration of concrete, and are used in large or difficult pours where partial setting before the pour is complete is undesirable.
- Air entrainers add and distribute tiny air bubbles within the concrete, which reduces damage during freeze-thaw cycles in winter, thus increasing the concrete's durability. However, entrained air has a disadvantage in that there is a trade-off with strength – each 1 per cent of air can result in 5 per cent reduction in compressive strength.
- Plasticizers increase the workability of 'fresh' concrete, allowing it to be placed more easily and compacted with less effort. Alternatively, plasticizers can be purposely employed to

Admixtures

As admixtures alter the properties and performance of concrete, they must always be used in accordance with manufacturer's instructions. They must be gauged very carefully at the mixing stage and from one batch to the next, so that each successive batch contains exactly the same amount as the previous one.

reduce the water content of a concrete; in these circumstances they are referred to as 'water reducers'. This improves the strength and durability characteristics of the concrete whilst still maintaining workability during placing.
- Pigments or colourizers can be used to change the colour of concrete, for aesthetic and design purposes. Most pigments come in a liquid form for adding to the mixing water.
- Corrosion inhibitors are used to minimize the corrosion of steel and steel reinforcement bars in concrete.
- Bonding agents are used to create a bond between old and new concrete.

CALCULATING QUANTITIES OF MATERIALS FOR CONCRETE

As a general 'rule of thumb', one cubic metre (1m^3) of compacted and finished concrete weighs 2400kg. It is easy, therefore, to calculate the quantity of concrete based on the weight of a concrete slab or foundation. From this, the quantity of each component part of the mix can be calculated based on the mix proportions used.

For example, for a foundation slab 10m long × 4m wide × 0.25m thick, using a concrete mix of 1:2:5 (cement:fine aggregate:coarse aggregate), the calculations would be as follows:

Volume of concrete = 10m x 4m × 0.25m = 10m^3 @ 1:2:5 (total of 8 component parts).

Weight of concrete = 2400kg/m^3 × 10m^3 = 24000kg.
Add an amount for waste (typically 10 per cent) = 24000kg × 10 per cent = 26400kg @ 1:2:5.

So the quantities are:

- Cement = (26400kg ÷ 8 parts) × 1 = 3300kg
- Fine agg' = (26400kg ÷ 8 parts) × 2 = 6600kg
- Coarse agg' = (26400kg ÷ 8 parts) × 5 = 16500kg

Water content is typically 50 per cent of the cement content by weight; 1kg of water = 1 litre, so the amount of water required = 3300kg × 50 per cent = 1650kg = 1650 litres.

Note: when calculating quantities by weight, there is no need to take account of volume shrinkage on mixing!

Concreting Checklist

Good quality concrete must:

- have the correct mix proportions;
- use well-graded aggregates;
- use clean and dry (whenever possible) aggregates;
- have the correct water:cement ratio;
- have adequate but not excessive workability for the circumstances;
- be properly and sufficiently mixed;
- not segregate during mixing or transportation;
- be well compacted;
- be adequately cured;
- be adequately protected from the elements;
- be free from shrinkage cracks;
- be durable and weather resistant;
- be of sufficient strength and fit for purpose.

CHAPTER 3

Mortar

Mortar is the 'chemical glue' and 'gap-filling adhesive' that holds masonry structures together. Its purpose is to even out any slight irregularities in brick shape and size and it has a direct influence on compressive strength, durability and resistance to rain penetration through brick walling. On this basis it is as fundamental to the construction of a wall as the bricks or blocks themselves.

In basic terms, mortar is a mixture of an aggregate (soft sand or builders' sand), a binder (usually Ordinary Portland cement) and water. Additives such as plasticizers and pigments can be added in order to achieve a particular performance objective. Water is added to form a paste with the cement, which covers the surface area of every sand particle in the mix, binding the particles together to form a solid mass when the mortar hardens. The sand is not physically altered in any way, but is firmly set into the hardened, rock-like cement paste.

REQUIREMENTS

From a bricklayer's point of view a mortar should be 'fatty' – in other words, it should handle well without being sticky, spread easily and set at the right pace to allow time to finish the joints in all weather conditions. If a mortar satisfies all these requirements it is said to have 'workability'. Harsh mortars with low workability slow down the bricklaying process and reduce a bricklayer's output. Picking up and spreading the mortar is more difficult, as is placing the cross-joints on the ends of bricks – harsh mortars tend to have poor adhesion and will fall away from the brick and/or trowel. Workability is much improved by the addition of plasticizers or lime to the mix. In achieving 'workability' a number of other factors must not be overlooked.

Retention of Plasticity

The length of time that mortar retains its plasticity is something of a trade-off. Plasticity should be retained long enough to allow the bricks to be laid, adjusted and a joint finish applied before the mortar dries out too much. This is more of a problem where very dry bricks are used, as they have a high suction rate. Alternatively, mortars that are excessively plasticized can be laid only to limited heights before the wall sinks and bed joints of the lower courses start to squeeze out.

Durability

The extent of a mortar's durability (in other words, its resistance to weather, frost and chemical attack), as determined by the mix proportions, must be suitable for the purpose for which it is used. For example, a comparatively weak mix may be suitable for an internal block partition but the same mix would weather and deteriorate quite quickly if used on an exposed boundary wall.

Good Bond with Bricks

Strong mortar mixes with a high cement content have a greater proclivity to drying shrinkage, which results in the mortar joints effectively shrinking away from the bricks. The fine cracks between joint and brick undermine structural strength and allow rain to penetrate, with the potential for frost damage during winter months. Some bricks, such as concrete or sand/lime bricks, also display high rates of drying shrinkage and such problems are exacerbated if they are laid in harsh, strong mortars.

Adequacy of Compressive Strength

Mortars that are overly strong will respond to differential movement with cracks, which may be few in number but they will be large, and usually through the bricks or blocks themselves rather than in the mortar joints. Weaker mortars on the other hand are capable of accommodating small amounts of structural movement and any cracking will be hairline in nature and concentrated in the joints. In the event that remedial work is required, raking out and re-pointing joints is a much easier and quicker task than the wholesale chopping-out and replacing of cracked bricks.

MORTAR MIX DESIGN

It is not possible for one mortar mix to provide all the required characteristics to the maximum extent. For example, great strength can be achieved using a 3:1 sand and cement mortar but this is at the expense of poor workability, for one thing. For a given set of circumstances, then, the ideal mortar mix has the following characteristics:

- it provides an appropriate balance between good workability and plasticity retention for the bricklayer;
- a good bond with the bricks or blocks being laid;
- it allows joints to be compressed and sealed against the driving rain and wind;
- its appearance is complementary to that of the bricks;
- it is sufficiently durable for its location or use and of adequate strength, but slightly less strong (and certainly no stronger) than the bricks or blocks being bedded. (This makes the choice of bricks a key determining factor in terms of mortar mix design.)

The final strength of the mortar is determined by the strength of the brick or block to be bedded in it. The mortar strength should roughly match that of the brick or block but it should not be greater, so that any cracking from movement occurs in the joints and not in the bricks, thus making remedial repairs easier.

Typically, these criteria can be most easily met by using a mix containing cement, hydraulic lime and sand, in proportions appropriate for its use.

Ordinary Portland Cement

Ordinary Portland cement is the most commonly used for bricklaying mortar. In order to ensure consistency of mortar colour, cement should be sourced from just one place for each building project.

Cement–Lime–Sand Mortar

Often referred to as 'compo', this mix contains a proportion of hydraulic lime for a number of reasons. Lime contributes little in terms of strength but it does act as a plasticizer to make the mortar more workable. Lime also has good water-retaining properties, which means that the setting time of the mortar is slowed, with the mortar staying workable and keeping its plasticity for longer. The presence of lime renders the set mortar somewhat more flexible than sand–cement mortar, meaning that differential structural movement is more easily accommodated. The addition of lime also enhances the aesthetics of mortar by giving it a more creamy appearance.

Hydraulic Lime

There are two types of lime: 'hydraulic' lime, which sets under water, and 'non-hydraulic' lime, which needs air in which to set. The construction industry commonly uses non-hydraulic lime, which is manufactured by heating pure limestone in a kiln to 1066 degrees centigrade. The resultant 'quicklime' is then mixed with water (in a process known as 'slaking') to make lime putty, or with less water to produce hydrated lime in the form of a fine, white powder that is delivered to site in 25-kg bags. Since lime is less dense than cement, a 25-kg bag of lime is significantly larger than a bag of cement of the same weight.

Bags of lime must be protected from damp before use. Preferably, bags should be stored clear of the ground on a wooden pallet in a well-ventilated, rain-proof shed. Bags of lime should be stacked flat, not more than five bags high. If stacking outside cannot be avoided, bags should be raised off the ground and adequately sheeted with polythene, ensuring a good overlap. It is vitally important to keep the bags dry.

Fig. 18 Size comparison of 25-kg bags of Ordinary Portland cement and hydraulic lime.

BELOW: **Fig. 19 Hydraulic lime.**

Before the widespread use of cement, mortar used in bricklaying was referred to as 'coarse stuff', a 3:1 ratio (aggregate to lime) mortar mix using non-hydraulic lime putty, well-graded sharp sand and soft sand. It is still available today from specialist suppliers and is suitable for repairs to historic buildings, blocklaying, bricklaying, stonemasonry, the re-pointing of brickwork, the repair and plastering of internal walls (as a backing or finishing coat), and the repair or rendering of external walls.

SAND

Sands and aggregates are natural products used in the production of mortar and concrete. They are usually excavated from locations where ancient river and sea beds were re-formed during the early periods of the earth's structure. The term 'aggregate' is a generic one used to describe gravels, stones and sharp sand, which are mixed with cement and water to make concrete.

Fig. 20 Soft red sand.

Fig. 21 Soft yellow sand.

Sand

In order to ensure consistency of mortar colour, always obtain sand from one source only for each building project.

The terms 'fine aggregate', 'concreting sand' or 'sharp sand' describe the natural sand, crushed stone or similar that passes through a 5mm sieve but tends to be coarser and not as well graded as the sands used in mortars. It is often referred to as 'fine aggregate' to distinguish it from soft sand. Being coarser than soft sand, it is not used for mortars as it produces an unworkable mix and causes difficulties with achieving a good finish when jointing. The sand used for mortar is referred to as 'soft sand' (or sometimes 'builders' sand'), as distinct from the fine aggregate (also known as 'sharp sand') used in the mixing of concrete.

Soft sand should be well graded, as poorly graded sand will require a greater amount of cement binder to fill the gaps between the grains. This will cause the mortar to shrink more, creating drying shrinkage cracks in the mortar joints and at the junction between the mortar joint and brick.

Like aggregates for concrete, sands can be purchased in 25-kg bags, 1-tonne 'dumpy bags' or loose by the lorryload. Soft red sand tends to produce mortar that is brown in colour whereas soft yellow sand produces grey mortar.

'Bulking' of Sand

Using sand that is not dry not only has implications for the water content of mortar mixes, which must be allowed for when mixing, but may also cause 'bulking', which is more serious.

Dry sand and saturated sand have about the same volume but a sample of damp sand will show a marked increase in volume. In damp sand, a film of water forms around each grain of sand, pushing it away from its neighbouring grains. This is known as 'bulking' and the volumetric increase can be as great as 30 per cent. When sand becomes saturated, the sheer volume of excess water present causes the surface tension round each grain to break down, preventing the bulking effect.

The use of damp sand can have an adverse effect on mortar mixes if they are being gauged by volume, because the mix will contain proportionately less sand in relation to cement than seems to be the case, resulting in a stronger mix than desired. Accordingly, it is best to use dry sand for mixing mortar.

Cleanliness of Sand

Impurities and contaminants in sand can cause staining within the finished mortar, or weaken it. The most simple test is to squeeze the sand between the fingers; if a stain remains, the sand should not be used.

The 'Field Settling Test' is a more scientific assessment that can be carried out on site to give a guide to the amount of silt in natural sand. Reasonable accuracy can be achieved with a straight-sided glass, vase or a 500g jam jar, plus a measuring tape to calculate the results:

1. Fill the glass or jar up to a height of about 50mm with a salt-water solution (one teaspoonful of salt to 750ml of water).
2. Pour in sand up to a level of 100mm.
3. Add more salt solution until the level rises to 150mm.
4. Shake the sand and salt solution well.
5. Stand the glass, vase, or jar on a level surface and tap it until the top of the sand is level.
6. Leave it to stand for three hours, during which a layer of silt will be clearly seen to form on top of the sand.
7. Measure the height of the silt layer and the height of the sand layer.

The silt content percentage is calculated by taking the height of the silt layer and dividing by the height of the sand, then multiplying the result by 100.

The silt content should be not more than 8 per cent. If it is more than 8 per cent, the sand should not be used, as excess silt or clay will inhibit a good bond between the cement and aggregates, causing weaknesses in the finished mix. The example in Fig 22 has a huge silt content, clearly more than the 8 per cent minimum, making this sand unsuitable for use in mortar.

For an even more accurate assessment, use a 250ml measuring cylinder and substitute volume

Fig. 22 Result of a field settling test carried out on soft sand.

for height when measuring out the ingredients for the test and for calculating the silt content.

MIXING MORTAR

Mixing Water

The essential criterion for mixing water is that the water should be clean, as any impurities will affect the strength of the finished mortar or result in staining. The term 'drinkable' (or 'potable') is often used to define the quality of clean and uncontaminated water fit for use in mortar.

> **Gauging of Mortar Mix Materials**
>
> In this context, 'gauging' is the term used to describe the measuring out of quantities of individual dry materials for a mortar mix. It is common practice on construction projects simply to load a mixer using a shovel, meaning there is no accurate control over the proportions of materials. This is, however, bad practice and the only way to achieve a degree of accuracy is to use batching by weight or by volume. The principles are the same as those applied when measuring out materials for concrete mixes. For more on the importance of good storage arrangements, grading of aggregates and gauging of materials, see Chapter 2.

Mortar sets as a result of the hydration process that takes place between the water and cement, which makes the cement set hard and binds the sand particles together. Only a small amount of water is actually required for the hydration process to take place, but enough water must be added to make the mortar workable. This will always exceed the amount needed for hydration of cement. In essence, water is added until the mix 'feels right' and meets the workability requirements of the bricklayer.

In reality, a typical cement–lime–sand mix of 1:1:6 will require approximately 7 litres of water for every 5kg of cement. A vastly excessive amount of water produces an overly workable mix, can lighten the colour of the finished mortar and result in a weakened mortar mix.

Due to the amount of water that is present, mixed mortar will be usable for a few hours, depending on weather conditions. When it does start to 'go off' it can be 'knocked up' by adding

Typical Cement–Lime–Sand Mortar Mixes

Engineering bricks and dense concrete blocks	*Bricks and blocks, average strength. Below DPC*	*Bricks and blocks, average strength. Above DPC*	*Lightweight blocks in internal walls*
1:¼:3	1:1:6	1:2:8	1:3:10

more water, but this is not good practice as the mortar will have begun to set by hydration and will already have attained some of its initial strength. Re-used mortar and re-diluted cement paste will not achieve the original strength capability when they finally set.

Mixing Mortar by Hand

Small quantities of mortar will often be mixed by hand. The area selected for mixing must be flat, hard, and free from debris and dust. A concrete floor or patio area is ideal but the following must be considered:

- Mortar will inevitably stain the mixing surface, even if it is hosed down immediately afterwards, so an area should be chosen where this does not matter.
- There should be enough space to move around easily, use the mixing shovel and turn the materials. Preferably the selected area should be close to where mixing materials are stored and/or where the mixed mortar will be needed.
- All batches should be mixed for the same amount of time, ensuring that all the sand is coated with cement paste and the mortar is a uniform colour, with all materials thoroughly integrated.

It is possible to acquire a bespoke large plastic mixing tray with raised edges from a builders' merchant, but it may be a problem to store later or an unnecessary expense if it is only to be used once or infrequently.

The following tools and equipment will be required for hand-mixing mortar:

- a wheelbarrow for transporting materials;
- a water butt, such as a plastic or metal drum or barrel, or alternatively access to an outside tap or hose;
- builders' plastic buckets, preferably three: one for water, one for cement and one for sand, since the latter may contain some moisture. An alternative to a bucket is a gauge box;
- a builder's shovel for mixing.

Measure out the correct proportions of materials and place the dry ingredients together on the mixing area. Using a shovel, turn the mix over into a single pile, ensuring good integration and uniform colour of the dry materials. Repeat the process by turning the mix back again. The main object of this is to distribute the cement evenly within the particles of sand. Repeat this process a third time, at which point the dry mix should be fully integrated.

Make a hole in the centre of the dry mix ready to receive the water from a bucket. It is vital to ensure that the water does not escape, taking cement with it and thus reducing the strength of the mix. The materials should be mixed with the water from the outside into the middle until all the water has gone. Then turn all the materials over on to the top of the heap and continue turning the whole mix over until it is fully mixed and workable. Use a shovel to place the mixed mortar into a barrow. Wet the inside of the barrow first to assist with tipping out the mix later.

Mixing Mortar by Machine

1. Set the mixer up according to the manufacturer's instructions. It is always a good idea to use polythene to sheet down under and around the mixer to minimize mess.
2. Ensure you have sufficient materials and small tools.
3. Start the mixer.

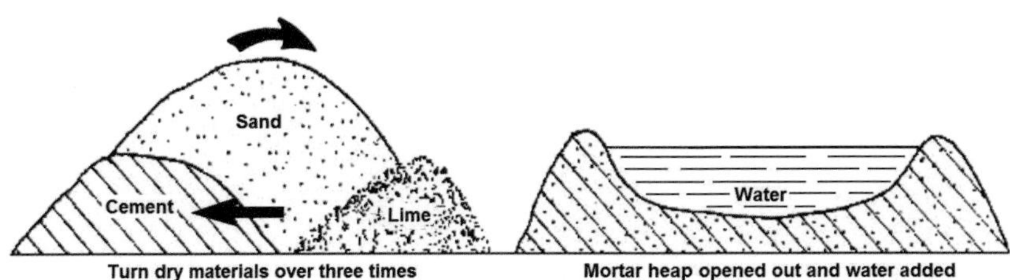

Fig. 23 Mixing mortar by hand.

Fig. 24 Ready-mix mortar delivered to site.

4. Add three-quarters of the water using a bucket.
5. Add three-quarters of the sand and three-quarters of the lime.
6. Gradually add the cement and allow to mix in.
7. Add the remainder of the sand, lime and water.
8. Allow the materials to mix for enough time to ensure a uniform colour and full integration.
9. Add more water as necessary to achieve the required workability.
10. Allow the materials to mix for a further 3–5 minutes – any longer, and too much air may be entrained into the mix, which will weaken the mortar. Timing should be taken from when all of the materials are in the barrel of the mixer and all batches should be mixed for the same length of time. Insufficient mixing time results in a non-uniform mix and poor workability.
11. Turn out the mix into the wheelbarrow, after making the inside wet.

> After completing the mixing, the mixing area and all tools and equipment should be thoroughly cleaned and tidied, otherwise cement paste will harden on the tools and will be difficult to remove later. This is particularly important if pigments have been used in the mix.

There is some difference of opinion as to the order in which materials should be added to the mixer. With regard to the order listed above, some take the view that unless the cement is added extremely carefully it is likely to 'ball' together and not be evenly distributed throughout the mix. This will lead to a weaker mix and long-term problems of staining.

An alternative method is to load the mixer with three-quarters of the water then gradually and slowly add all of the cement, to achieve a thin paste with no lumps in it. The remaining dry materials and water are then added.

There is one point on which everyone agrees, however: if all the dry materials are added first, before adding any water, much of the cement will cling to the inside of the drum, resulting in a lean, weaker mortar mix.

READY-TO-USE MORTARS

Ready-to-use mortar is factory-produced off-site and then delivered in large plastic skips. Ready-to-use mortars contain retarding additives, which slow the hydration and setting process, so that the working life of such mortars varies from a minimum of 36 hours up to 72 hours, depending on the nature of the agents that have been added. Factory production has the advantage that it ensures consistent quality from one delivery to the next.

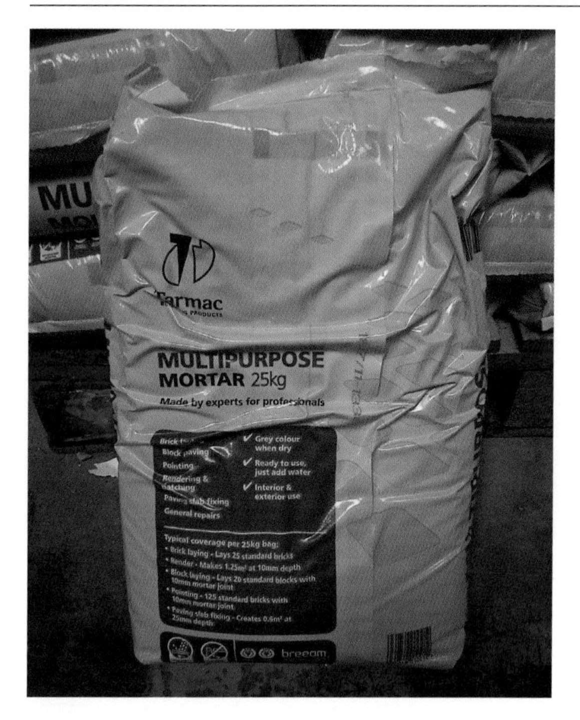

Fig. 25 Pre-bagged dry mortar (25kg).

PRE-BAGGED DRY MORTAR

Bags of pre-bagged dry mortar are available from builders' merchants in 25-kg bags and have a number of advantages:

- Batching of dry materials takes place at the factory so mix proportions are consistent.
- Delivery and/or stockpiling of loose materials can be avoided.
- Bagged materials can be offloaded more quickly and cleanly than loose materials.
- Bagged dry mortar is quite an economic method on small jobs and using it also reduces waste. This is, however, unlikely to be the case on large projects.

Smaller-sized bags (5kg for example) are available from DIY outlets but tend to be sand and cement mixes, with all the disadvantages associated with a lack of lime.

MORTAR ADMIXTURES

Lime acts as a 'plasticizer' but there are other chemical additives (in liquid and powder forms) that can be used in the mixing water that do the same job. These include accelerators to make mixes set more quickly (useful in frosty conditions); 'retarders' to slow down the setting process, to give more time to work the mortar and give a harder set; and 'pigments' or 'colourizers', to give a range of colours to mortars.

As admixtures alter the properties and performance of mortar, they must always be used in accordance with manufacturer's instructions. They must also be gauged very carefully at the mixing stage and from one batch to the next, to ensure that each successive batch contains exactly the same amount as the previous one.

Mortar Checklist

Good quality mortar must:

- contain the correct mix proportions for the strength and durability required;
- be no stronger than the bricks or blocks being bedded in it;
- contain well-graded, dry, clean sand;
- be of consistent mix proportions from one batch to the next to ensure uniform colour and strength;
- contain consistent quantities of admixtures (where applicable) from one batch to the next;
- contain cement from the same source to ensure uniformity of colour;
- contain sand from the same source to ensure uniformity of colour;
- be properly and adequately mixed;
- not contain excessive water but have adequate workability.

CHAPTER 4

Bricks and Blocks

There are thousands of bricks to choose from, in a wide variety of colours and surface finishes, and offering many different performance characteristics for different applications. In the main, however, bricks are a standard size of 215mm long × 102.5mm wide × 65mm high. These sizes have not been arrived at randomly but are carefully designed so that bricks coordinate together with an in-built allowance for a 10mm mortar joint, allowing them to be bonded in a walling situation.

Brick Manufacture

The vast majority of bricks used in construction are made from the naturally occurring raw material of clay. A small proportion are made from calcium silicate (in other words, sand/lime) or concrete, but this chapter will focus on those manufactured from clay.

In the manufacture of bricks, selected clays are prepared, moulded into shape and burnt under a

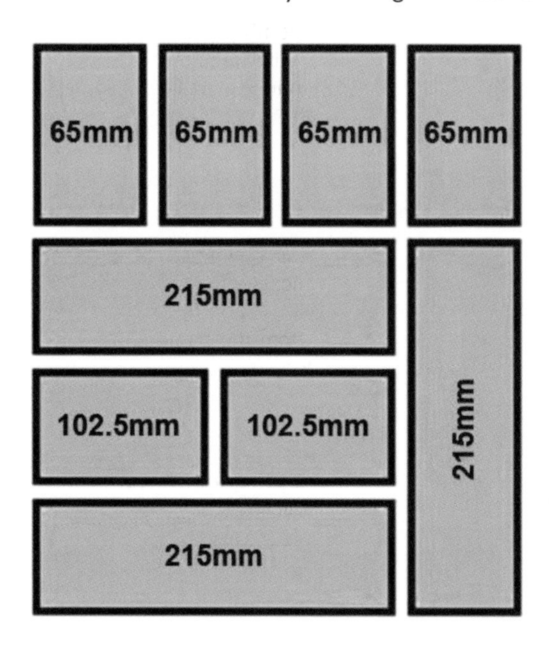

Fig. 26 Co-ordinating sizes of bricks with an allowance of 10mm for mortar joints.

Fig. 27 Solid brick (engineering).

Fig. 28 Perforated blocks (engineering).

Brick Terms

A variety of terms and terminology will be used here; *see* Fig 29 for some definition and clarity.

The question of what constitutes a 'solid' brick as opposed to a 'perforated' brick is not clear-cut. A 'solid' brick can still be defined as such, even when it has holes or voids, provided they do not exceed 25 per cent of the brick's total volume. Where holes exceed 25 per cent of the total volume of the brick, and where the holes are considered to be small, the brick is defined as 'perforated'. Where the holes exceed 25 per cent of the total volume of the brick and are large, the brick is defined as 'hollow'. Finally, 'cellular' bricks have holes that are only open at one end and exceed 20 per cent of the total volume of the brick. A 'frogged' brick where the frog is very deep could well be considered to be a cellular brick.

Despite these 'grey areas' in definition, for the purposes of this book, reference to 'solid' is intended to mean bricks with no holes at all (Fig 27), while 'perforated' refers to bricks or blocks with any holes, regardless of size, that pass all the way through (Fig 28). Any 'frogged' brick will be referred to as such, regardless of the depth of the indentation (Fig 30).

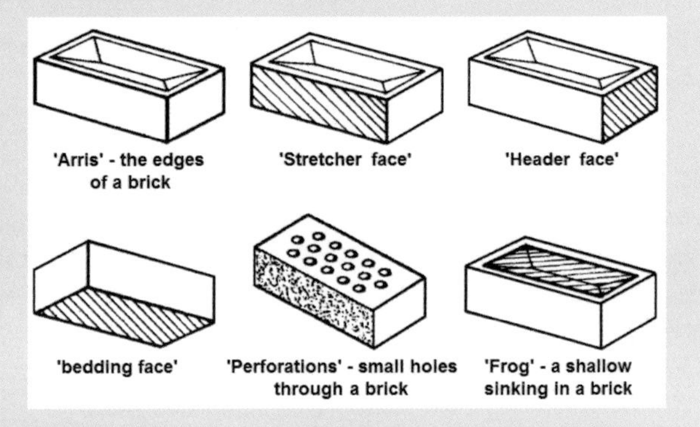

Fig. 29 Brick terms and terminology.

variety of manufacturing processes, each of which will produce a brick that possesses certain qualities or properties. Various colours of brick can be made by using iron oxides, iron sulphides, and other additives. The mineral content of the clay raw material will also have an effect on the physical properties, colour and hardness of the finished brick.

All clays used for brick manufacture have one common characteristic: they must be capable of being finely ground by machine and then mixed with water, so that they may be moulded, or 'pugged'.

Fig. 30 Frogged brick (facing).

After excavation from clay pits, the clay must first be 'weathered' to wash out all of the impurities and soluble salts, which could later lead to efflorescence on the face of the finished brickwork. The most basic method is to leave piles of clay outdoors and open to the elements during the winter so that the rain simply washes out the soluble salts. Other clays are passed through a washmill first before being stored in large open storage areas.

The washed clay then undergoes a number of grinding processes (usually three) until the raw material is reduced to a particle size of 1–2mm, at which point water is added in order to enable the clay to be moulded into brick shapes by one of the following three basic methods.

Wire-Cut Extruded Bricks

The clay is mixed with water to a fairly stiff consistency and loaded into an extruding machine, where it is forced under pressure out through a die. The die is machined to a shape and size that is larger than the finished size of the brick, to take account of the extent to which clay will shrink during the later drying and firing processes. A typical size for the die is 240mm × 125mm.

The clay emerges from the die as a continuous, smooth, brick-shaped column. At this point a decorative finish may be applied to the faces of the brick – a very taut wire is used to slice a thin sliver off the top and both sides in order to produce a 'wire-drag' pattern. An alternative finish can be applied by way of textured rollers, or a 'sandfaced' finish can be achieved by blasting the column of clay with sand under high pressure.

The column of clay is then cut into single bricks (typically 75mm thick) by means of wires attached to a framework. The wire-cut bricks are then palletized ready for the drying process. Wire-cut extruded bricks are generally solid or perforated and are further characterized by sharp edges or arrises.

Soft-Mud Moulded Bricks

In 'soft-mud moulding' the clay is mixed with water to a fairly soft consistency and a 'clot' of the soft clay is thrown by hand into a mould box. The mould is oversized to allow for shrinkage of the brick during the later drying and firing processes. The clay is prevented from sticking to the inside of the mould by way of a releasing agent such as sand, oil or water. The excess clay is struck off from the top of the mould box and the bricks are then turned out. Using sand as a releasing agent will produce a brick with a 'sandy' texture, whereas using oil or water will produce a smooth finish. Making bricks by hand in this way is clearly slow and labour-intensive and they are, therefore, expensive to make in large quantities.

For large quantities of standard-sized bricks the process of 'soft-mud moulding' by hand can be replicated much faster and on a grander scale using large automated machines. These employ banks of mould boxes on a continuous circuit, with the boxes being washed, sanded, filled with clots of clay and struck off level, and the formed bricks being turned out in a matter of seconds. Again, using oil or water as a releasing agent will result in a smoother finish.

Soft-mud moulded bricks are either solid or have an indentation called a frog. They are characterized by having edges that are not sharp, being somewhat irregular in shape and lacking uniformity of size. These characteristics tend to be more noticeable in the hand-made varieties. In addition, the action of throwing by hand or mechanically dropping the clay into the moulds means a crease is achieved in the face of the brick, which creates a decorative effect that varies from brick to brick.

Machine-Pressed Bricks

The process for machine-pressing bricks ('soft-mud pressing') is an extension of the mechanized process of 'soft-mud moulding', with the soft clay being mechanically pressed into the mould under uniform pressure rather than simply dropped in. Greater quantities of clay under higher pressure can be used to make heavier, denser and more weather-resistant bricks. Being made in a mould, such bricks are either solid or 'frogged' and, having been machine-pressed, have sharp edges, are regular in shape and are uniform in size.

Drying Bricks

Before the 'green' clay bricks can be fired in a kiln, the maximum amount of moisture must be removed, otherwise the bricks will explode during the firing process. Drying must be done in such a way that the moisture is evenly removed from the inside out. If the outer faces of the brick are allowed to dry first then water from the inside of the brick cannot escape and remains trapped. Any trapped moisture will be forced out by the high temperatures in the kiln during firing and cracking of the brick is likely to result.

To prevent this, brick-drying chambers are maintained at temperatures of about 80 to 120 degrees centigrade within a very humid atmosphere, which keeps the exterior of the bricks moist. The bricks shrink during the drying process (which can last up to 40 hours) as the clay particles come together, whilst temperature and humidity are closely monitored to minimize surface cracking.

Upon completion of the drying process, the bricks have no weather-resistant qualities but they do have sufficient strength to be stacked and transported on kiln cars ready for firing.

Firing or Burning Bricks

Firing temperatures are often quite critical and will vary considerably between different clay types but are generally in the range of 900–1300 degrees centigrade. The length of firing is another key factor that has an impact on the properties of the finished brick.

The firing process takes place in three stages. First, preheating ensures total dryness of the bricks and raises the brick temperature. Next, a fuel such as natural gas, oil or coal is used to raise and maintain the temperature to the required firing level over a few hours. Finally, cold air is drawn into the kiln to cool the bricks slowly, ready for sorting and packing.

Bricks undergo a physical change during the firing process. Clay particles and impurities are fused together to produce a hard, durable and weather-resistant finished product. This process is known as 'vitrification' and is accompanied by further shrinkage and a colour change.

General Classification of Bricks

There are literally thousands of different bricks, with a multitude of colours, finishes, characteristics and practical applications. To address all of the differences would be a massive task but there follows a broad overview of how bricks can be differentiated and generally classified in terms of variety, quality and type. Applying such broad criteria will enable bricks to be generally specified for different applications and uses.

A bricklayer will normally classify bricks according to three varieties: common bricks, facing bricks and engineering bricks.

Common Bricks

Common bricks are fired to a temperature of between 950 and 115 degrees centigrade and are cheaper simply because less fuel is expended in firing them at a lower temperature. They have no decorative faces and are general building bricks that can vary greatly in quality. 'Commons' are normally used for unseen work or below ground (if sufficiently durable).

Fig. 31 Common brick (frogged).

Fig. 32 Facing brick (perforated).

Facing Bricks

Facing bricks are specially made to give an attractive appearance and are used specifically for this purpose and generally above horizontal DPC level. Chemical additives are used to give the bricks a pleasing colour and textures are applied during the manufacturing process. 'Facings' have a decorative finish to both header faces and one stretcher face. The remaining plain stretcher face is referred to as the 'common face'.

Engineering Bricks

Engineering bricks are fired to a high temperature of between 1200 and 1300 degrees centigrade until they are almost molten. The high firing temperature prevents the formation of air pockets or pores in the bricks and they are, therefore, dense, strong and non-porous, which makes them ideal for use in retaining walls, below DPC level, below ground, inspection chambers, copings to the top of walls, DPCs (when in solid form), and so on. 'Engineerings' generally weigh between 3.5 and 4.5kg each. Although they can be of 'facing' variety, most are smooth-faced, which means that they might be defined simultaneously as 'common', albeit of much better quality.

Fig. 33 Engineering bricks (perforated).

Quality of Bricks

The three usual varieties of brick can vary in quality to some degree within each classification, which means that a greater degree of clarity is required in order correctly to specify a brick for a particular purpose. To this end, there are five generally accepted levels of quality for bricks.

The first three levels are concerned with the durability and performance of a brick:

Internal Quality
Internal-quality bricks are not suitable for external work due to a lack of durability when exposed to the weather. Some commons are of internal quality as are some facing bricks.

Ordinary Quality
Ordinary-quality bricks are those that are suitable for most external uses; however, some may not stand up to very exposed conditions or below horizontal DPC where they will be wet and susceptible to frost damage. Most facing bricks fall into the category of ordinary quality, as do many commons.

Special Quality
The term 'special quality' comprises bricks that are durable even when used in conditions of extreme exposure where the structure may be saturated. This makes such bricks ideal in retaining walls, below DPC level, below ground, inspection chambers, copings to the top of walls, DPCs (when in solid form), and so on. Engineering bricks usually attain this level of quality as do some facing bricks. It must be remembered that, because many engineering bricks are smooth-faced, they can also be simultaneously classified as commons - on this basis it could also be argued that many commons achieve a quality defined as 'special'.

The remaining two levels of quality are more a measure of the appearance, colour and shape of a brick as opposed to durability and performance.

'Seconds'
Bricks that have become deformed and/or miscoloured during manufacture are known as 'seconds' and are frequently sold at a lower price than the 'perfect' equivalent. Seconds are unlikely to be used for facing work and certainly not for facing work of high quality.

'Selected'
On 'selected' bricks the quality falls somewhere between 'perfect' and 'second' – for example, the brick may suffer from minor discoloration only. The term 'selected' is quite subjective and usually used at the discretion of the supplier.

Variations in Brick Size

Bricks are man-made components manufactured from a naturally occurring material, so they will never be precisely the same from one brick to the next. In addition, different parts of the kiln will vary in temperature, and this can also affect the extent to which the brick shrinks during firing. It is common, therefore, for a dimensional variance to occur within a batch of bricks, with length being

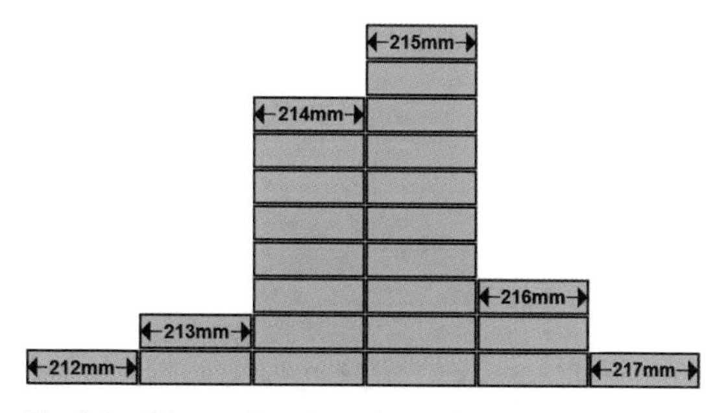

Fig. 34 Dimensional variance in a sample of twenty-five bricks extracted from a larger batch.

the most significantly affected. This variance in size is a key factor in brick quality. In the representative sample of twenty-five bricks in Fig 34, the vast majority of bricks conformed to the required size of 215mm or just below. Larger size variations and/or a greater number of bricks that stray too far from being 215mm will cause practical problems for the bricklayer, particularly, among other things, in maintaining a consistent cross-joint thickness. Bricks that display such excessive variations could end up being classified as 'seconds'.

Special-Shaped Bricks

Modern brickwork is often regarded as being rather dull and repetitive, but careful selection and use of materials, allied to some skilful bricklaying, can elevate it to more of an art form, giving designers and architects enormous scope for freedom of expression.

Most brick manufacturers offer a standard range of 'special' bricks to complement and match many of their facing and engineering bricks. Because the base material for bricks – clay – is so easily moulded into different shapes, many more non-standard 'specials' are available to order.

CONCRETE BLOCKS

Concrete blocks have been used in the construction industry since the 1930s for a variety of purposes including cavity walls, internal partitions and load-bearing walls, although they did not commonly replace brick for the internal skin of cavity walls for another 30 years or so. A block is defined by BS 2028 as a walling unit that exceeds the length, width or height of a brick. In other words, a block is any walling unit that is bigger in at least one dimension than a brick measuring 215mm × 102.5mm × 65mm. That said, the height of a block must not exceed its length or six times its thickness. Typically the dimensions of blocks are 440mm long × 215mm high × 100mm wide (although different widths are available for different purposes).

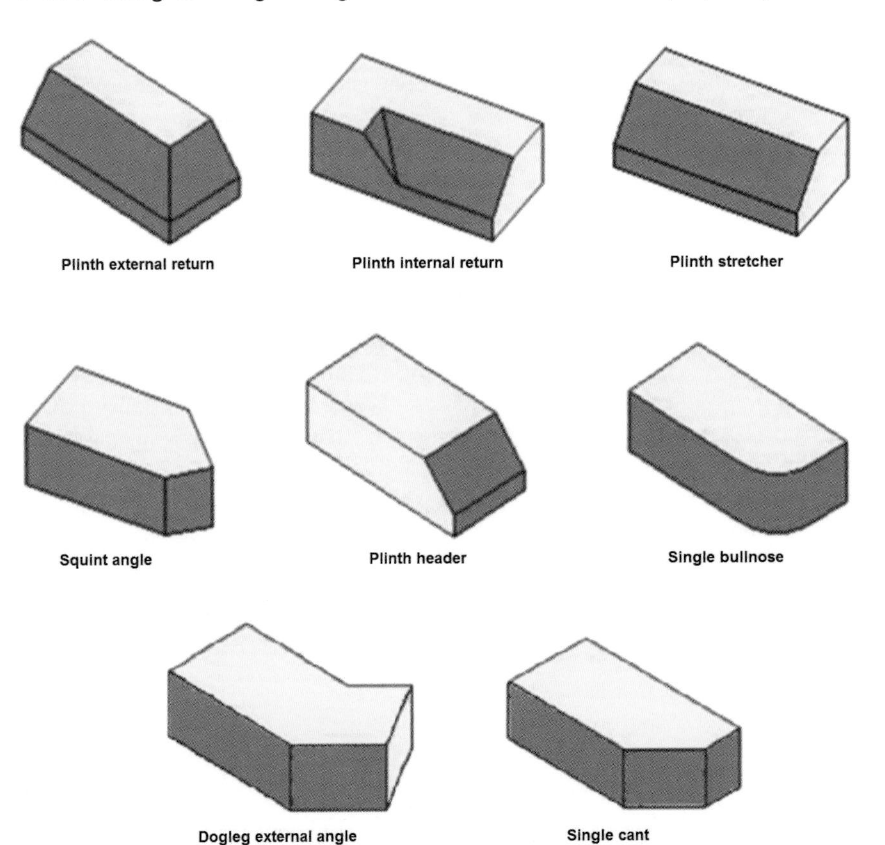

Fig. 35 Examples of special-shaped bricks.

Fig. 36 Concrete blocks.

The use of blocks has increased greatly since the late 1950s not only as a quick method of producing internal walls and partitions but also as a facing material in their own right. It is not uncommon to find neatly jointed, fair-faced blockwork as the finished wall surface inside a garage although the most common use is for the internal skins of cavity walls. Blocks have a number of advantages over bricks, the most obvious being that of productivity. By way of an example, laying one block with a 'coordinating size' (in other words, allowing for 10mm mortar joints) of 450mm × 225mm × 100mm is the equivalent of laying six bricks.

Blocks are generally divided into two types based on their material composition.

Dense Concrete Blocks

Dense concrete blocks are usually manufactured from cement, fine aggregate and coarse aggregate moulded under pressure, and are produced in a range of crushing strengths (measured in Newtons/mm^2) and widths up to 300mm ('trench blocks'). Being dense (in the region of 2200kg to 2400kg/m^3), they have poor thermal insulation qualities, so they tend to get used for load-bearing partitions, structural walls, sub-structure work (below ground) and party walls. When used externally above ground, dense concrete blocks are usually rendered due to their tendency to absorb a lot of water. Any walls built in dense blocks that will be rendered or plastered should be allowed to dry out thoroughly first, as they will shrink on drying and could crack any plaster or render that has been put on top prematurely.

Their high density also means they are very heavy as constructional units (up to nearly 21kg in some cases for a 100mm block), which makes them difficult and tiresome to handle on site. Dense concrete blocks should be laid in mortars of average strength – 1:1:6 or 1:2:9 (cement:lime:sand) – although slightly stronger mixes are recommended below ground.

Lightweight Concrete Blocks

Lightweight concrete blocks have been available for as long as dense concrete blocks and were initially made to be lighter and easier to handle than their dense counterparts. They were manufactured using a wide range of lighter aggregate materials instead of fine and coarse natural aggregates. The aggregate materials – fuel ash, furnace clinker, coke (for 'breeze blocks'), and slag from iron and steel blast furnaces – were by-products of industrial processes, so local availability would determine how the blocks were made in different areas of the UK. A number of these aggregates, such as blast-furnace slag, contained significant quantities of hydraulic lime meaning that smaller amounts of cement were required to make lightweight blocks. Depending on the aggregate being used and the strength of block required, the aggregate:cement ratio would typically range from 6:1 to 8:1. With the reduced cement content and the use of industrial by-products, lightweight blocks are much cheaper to produce.

Depending on their density and strength, lightweight blocks are used for load-bearing and non-load-bearing walls, partitions and cavity walls. Despite being referred to as 'lightweight', they are still reasonably heavy (around 10kg).

Aerated Concrete Blocks

Since the late 1960s, much greater emphasis has been placed on the thermal insulation qualities of the lighter-weight blocks than on their ease of handling. As a result, many of the older-style blocks have been superseded by blocks made of aerated concrete ('aircrete'). Aerated concrete blocks are classified as a lightweight block but the unique manufacturing process, the properties of the finished block and their extensive use make them worthy of separate consideration.

Fig. 37 Aerated concrete block.

Aerated concrete blocks have a very low density of 475kg/m^3, making individual blocks very lightweight indeed (a standard solid 100mm block weighs around 4.5kg).

Manufacture of Aerated Concrete Blocks

Aerated blocks are made from cement, lime, sand, pulverized fuel ash (PFA, a by-product from power stations) and water. The PFA, sand and water are mixed together into a slurry which is then heated and mixed with the cement and lime. Next, a small quantity of aluminium powder is evenly dispersed within the mixture, chemically reacting with the other constituents to form millions of tiny bubbles of hydrogen. The mixture is then poured into elongated moulds.

The hydrogen subsequently diffuses out from the material as it sets, to be replaced by air, creating an internal micro-cellular structure of millions of air pockets (see Fig 38). When the concrete has partially set, the long strips of aerated concrete are wire-cut into blocks of the correct size. The cut blocks are then transferred to a high-pressure autoclave for steam-curing, during which calcium silicates are formed, which permanently bind all the ingredients together into a solid mass.

Aerated concrete blocks are more expensive than more traditional concrete blocks (by approximately 50 per cent) and break easily if handled carelessly, but they have numerous advantages:

- they are easy and light to handle;
- they facilitate faster construction;
- they have a high ratio of strength to density;
- they are available in a variety of thicknesses, from 75mm to 230mm;
- they can be cut, drilled, chased and fixed to easily;
- they offer good heat insulation and low thermal conductivity;
- their sound transmittance is low;
- their fire resistance is good;
- their frost resistance good;
- their sulphate-resistant properties are good;
- the load on foundations is reduced.

Within their wide range of aerated blocks, manufacturers are always 'tinkering' with the constituents and manufacturing process to produce blocks that are enhanced in certain key areas such as strength, thermal transmittance, sound insulation or fire resistance; the last two criteria are particularly important for party walls.

Applications of Aerated Concrete Blocks

Due to the large number of benefits offered by

Fig. 38 Inside an aerated concrete block.

aerated concrete blocks, they are put to a wide range of uses and applications, with a variety of blocks available to suit different purposes

Standard blocks are available in a range of crushing strengths – 2.8N/mm^2, 3.5N/mm^2, 7.0N/mm^2 and 8.4N/mm^2 – and are used for internal partitions (load-bearing and non-load-bearing), internal skins of cavity walls, external skins of cavity walls (when rendered), external solid boundary walls (provided they are rendered or clad for weather resistance) beam and block floors, and sub-structure work below ground level. The higher-strength blocks tend to get used where greater strength is required, such as for two- and three-storey construction. 10.4N/mm^2 blocks are available for four-storey construction

Rendering Aerated Blocks

Aerated concrete blocks have a pattern on their surface which provides a key for plaster or render. However, the blockwork should also have its mortar joints raked out to a depth of 10mm to provide an additional key, and the blockwork surface dampened to provide additional suction prior to being rendered.

Wide blocks are available for use as 'trench-blocks' below ground, which eliminates the need for two skins of a cavity wall, plus cavity in-fill, to be built from the foundation to ground level. This greatly speeds up construction. Construction speed is also enhanced by some manufacturers of trenchblocks incorporating a tongue and groove moulding into the blocks, which enables them to essentially 'slot together'.

While aerated concrete blocks will receive paint and tiles directly, there are a number of block types available with a smooth, paint-grade surface. These are suitable where a higher standard of internal decorative finish is required without plastering.

Mortar mixes should not be too strong as aerated blocks do not accommodate movement very well. The recommended mortar mix, however, is much the same as for dense concrete blocks: cement:lime:sand at 1:1:6 or 1:2:9. Below ground level a slightly stronger mortar should be used, such as 1:½:4.

CHAPTER 5

Simple Foundations and Bases

FOUNDATION DESIGN

Loads and Bearing Capacity

The foundation for a building or wall is the part of the structure that is in direct contact with the ground. A structure will not stand without an adequate foundation. Its purpose is to transfer loads safely into the ground without excessive movement, settlement, or damage to the structure, throughout its life and during the initial construction process. The term 'load' has a number of meanings:

- The 'dead load' is the constant weight of the structure itself and all its component parts.
- 'Imposed loads' or 'live loads' are those loads that are created by the weight of people, vehicles, furniture and other objects placed in or on the structure. Inevitably, this type of loading can vary greatly.
- Other loads include the forces and stresses produced by wind and snow, which must be allowed for in the design of the foundation.

Foundation design depends largely on the type of sub-soil on which the structure is to be built. The soil must be capable of resisting the downward loads from the structure (including the foundations themselves) by way of equal and opposite upward forces. The soil, therefore, needs to have sufficient compressive strength or bearing capacity to prevent settlement and contraction.

Bearing capacity, which is a measure of how much pressure can be supported, varies from one soil type to another. Since Unit Pressure equals Force divided by Area, enlarging the plan size of the foundation reduces the pressure on the soil. A soil with a low bearing capacity will need foundations with a greater surface area. Alternatively, it may be necessary to excavate down to significant depth until a sub-soil of sufficient bearing capacity is found.

All soils have a tendency for movement due to changes in temperature and water content. In particular, cohesive, shrinkable soils such as clay can swell and shrink quite violently as their water content increases or decreases, and it has been known for the sub-structure of a building to be crushed. Non-cohesive soils, such as sandy ones, can be affected by surface water eroding particles or by freezing ('frost heave') in winter.

The depth of ground cover must be sufficient to prevent the foundation being subjected to 'frost

Building Regulations

The requirements of the Building Regulations, in the main, apply only to the construction of buildings. Consequently, when designing a foundation for a free-standing boundary or garden wall, with loadings that are significantly smaller, the excavation need be no deeper than is necessary to provide a soil with sufficient bearing capacity. This may actually be quite close to finished ground level. That said, however, the foundation must be deep enough to resist the effects of ground movement caused by changes in temperature and moisture content. For example, ground movement caused by drying shrinkage and expansion on becoming wet is particularly prevalent in clay sub-soils.

heave', so the foundation must be below the point at which frost can penetrate and expand, causing damage. Current Building Regulations require simple strip foundations to be located 1000mm below finished ground level.

Effect of Trees on Foundations

Generally, the ground level will rise and fall due to normal seasonal changes in water content but the presence of trees will exacerbate any problems, as they can draw a considerable amount of water out of the soil. It is an issue especially in clay sub-soils, which have the greatest tendency to shrink and swell with changes in water content. During dry summer months, trees will continue to draw increasing amounts of water from the ground causing a clay sub-soil to shrink further. Around a tree it is possible for the ground level to rise and/or fall by up to 40mm between winter and summer.

One option is to chop down the tree so that its roots are no longer draining water from the ground, but this can result in a clay sub-soil swelling by as much as 150mm. For this reason, the ground must be given an extended period of time to stabilize following the removal of trees. Due to the ground instability that results from the presence (or removal) of trees, and its potentially negative effects on foundations, it is generally accepted that foundations, and the structure thereon, should be kept well away from trees. As a 'rule of thumb' the distance should be at least equal to the height of an individual mature tree, or its anticipated height at maturity. Where several trees are located nearby, this distance could be increased to the height of the tallest of the group of trees.

The presence of dead trees near to proposed building work cannot be ignored as eventually they will rot below ground. This results in hollows and depressions forming in the ground which can have a negative effect of the bearing capacity of the ground near by.

> **Tree Roots**
>
> The physical presence of a tree can also result in structural damage to foundations and walls caused directly by root growth. In some cases, walls built too close to trees can have their foundation lifted up or even cracked, resulting in structural failure.

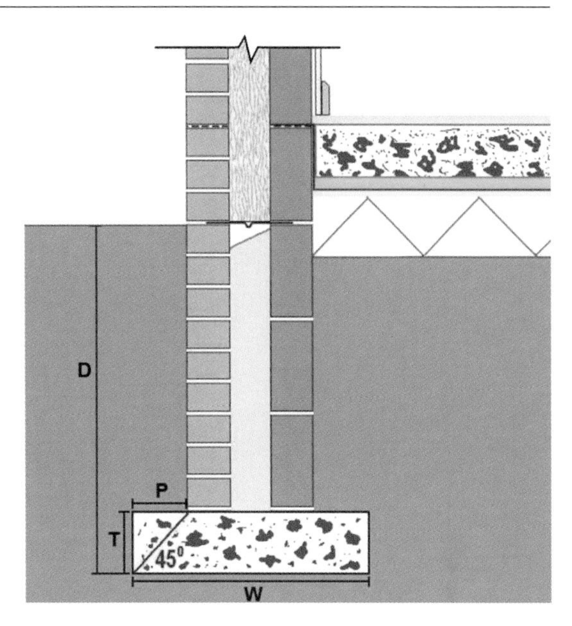

Fig. 39 Concrete strip foundation.

STRIP FOUNDATIONS

Traditional Strip Foundation

The simplest form of foundation is a mass-concrete strip positioned under free-standing walls or the walls of a building (*see* Fig 39). The minimum depth of ground cover for the foundation must be sufficient for the foundation not to be affected by ground movement or frost heave. This generally means excavating foundation trenches to a depth ('D') of 1000mm. Firmer sub-soils that are less susceptible to movement can have shallower foundation depths but this will depend on individual site conditions.

The thickness of the foundation ('T') must be a minimum of 150mm.

The width of the foundation ('W') depends on the bearing capacity of the soil and can be as little as the width of the wall (on rock sub-soils) or as much as 850mm. Typically, however, strip foundations are 650mm wide so as to provide working space in the trench.

Minimum Width of Strip Foundations

Type of sub-soil	*Condition of sub-soil*	*Physical characteristics*	*Minimum width in mm for total load in kN/m of load-bearing wall of not more than 20kN/m; 30kN/m; 40kN/m; 50kN/m; 60kN/m; 70kN/m*
Rock	Not inferior to sandstone, limestone or firm chalk	Requires at least a pneumatic or other mechanically operated pick for excavation of wall	In each case equal to width
Gravel	Compact	Requires pick for excavation. Wooden peg, 50mm square, hard to drive in beyond 150mm	250; 300; 400; 500; 600; 650
Sand	Compact	Requires pick for excavation. Wooden peg, 50mm square, hard to drive in beyond 150mm	250; 300; 400; 500; 600; 650
Clay	Stiff	Cannot be moulded with fingers, requires pick/pneumatic/mechanical spade for removal	250; 300; 400; 500; 600; 650
Sandy clay	Stiff	Cannot be moulded with fingers, requires pick/pneumatic/mechanical spade for removal	250; 300; 400; 500; 600; 650
Clay	Firm	Can be moulded by substantial pressure with the fingers and hand-dug with a spade	300; 350; 450; 600; 750; 850
Sandy clay	Firm	Can be moulded by substantial pressure with the fingers and hand-dug with a spade	300; 350; 450; 600; 750; 850
Sand	Loose	Can be excavated with a spade. Wooden peg 50mm square can be easily driven in	400; 600; must be expertly designed
Silt/ Sand	Loose	Can be excavated with a spade. Wooden peg 50mm square can be easily driven in	400; 600; must be expertly designed
Clay/ Sand	Loose	Can be excavated with a spade. Wooden peg 50mm square can be easily driven in	400; 600; must be expertly designed
Silt	Soft	Fairly easily moulded in the fingers and readily excavated	450; 600; must be expertly designed
Clay	Soft	Fairly easily moulded in the fingers and readily excavated	450; 600; must be expertly designed
Sandy clay	Soft	Fairly easily moulded in the fingers and readily excavated	450; 600; must be expertly designed
Silt/ clay	Soft	Fairly easily moulded in the fingers and readily excavated	450; 600; must be expertly designed
Silt	Very soft	Natural sample, in winter conditions, exudes between fingers when squeezed in fist	600; 850; must be expertly designed
Clay	Very soft	Natural sample, in winter conditions, exudes between fingers when squeezed in fist	600; 850; must be expertly designed
Sandy clay	Very soft	Natural sample, in winter conditions, exudes between fingers when squeezed in fist	600; 850; must be expertly designed
Silt/ clay	Very soft	Natural sample, in winter conditions, exudes between fingers when squeezed in fist	600; 850; must be expertly designed

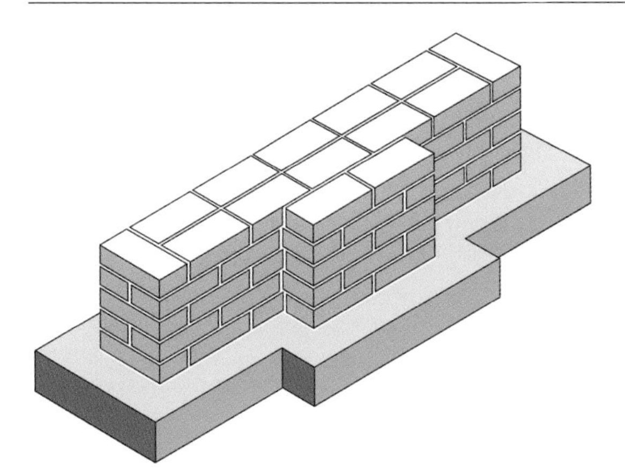

Fig. 40 Strip foundation with projection beneath and attached pier.

The projection either side from the wall face ('P') must generally not be greater than 'T'. This is because loads are transmitted through the foundation at 45 degrees so, if 'P' exceeds 'T', any cracking of the concrete strip will result in a reduction of the width and effective area of the foundation. Accordingly, wider foundations with greater projection ('P'), employed to counter weak sub-soils, will need to have increased thickness ('T') in order to maintain the 45-degree angle of load transfer.

The 'Minimum Width of Strip Foundations' table provides guidance as to the width of strip foundations appropriate for different soil types and conditions.

Where a wall or structure includes a projection (for example, an attached pier), the strip foundation must extend beyond the faces of that pier, or any buttress or chimney forming part of a wall, by at least as much as it projects beyond the face of the main part of the wall (see Fig 40).

Stepped Strip Foundation

Sloping construction sites have implications for the design of foundations. For example, if trenches for a strip foundation were excavated on a steeply sloping site by digging down to a depth of 1000mm at the low end of the site, by the time excavation to the same depth was finished at the high end of the site, the trenches could be many metres deep. Obviously, excessive or unnecessary excavation is expensive and time-consuming so it is minimized on sloping sites by stepping the foundation up the slope (*see* Fig 41), at the same time maintaining the required amount of ground cover.

At each step, the upper level of a stepped foundation overlaps the lower level by twice the thickness of the foundation or by 300mm, whichever is the greater. In addition, the height of the step must not exceed the thickness of the foundation. From a practical point of view and for convenience, the height of the steps is often purposely constructed to be a multiple of 75mm, to match the gauge of the brickwork and to avoid the need for any 'split courses'. Finally, it should be noted that the horizontal DPC of free-standing walls constructed on sloping sites will step down (*see* Fig 41), in order to follow generally the line of finished ground level.

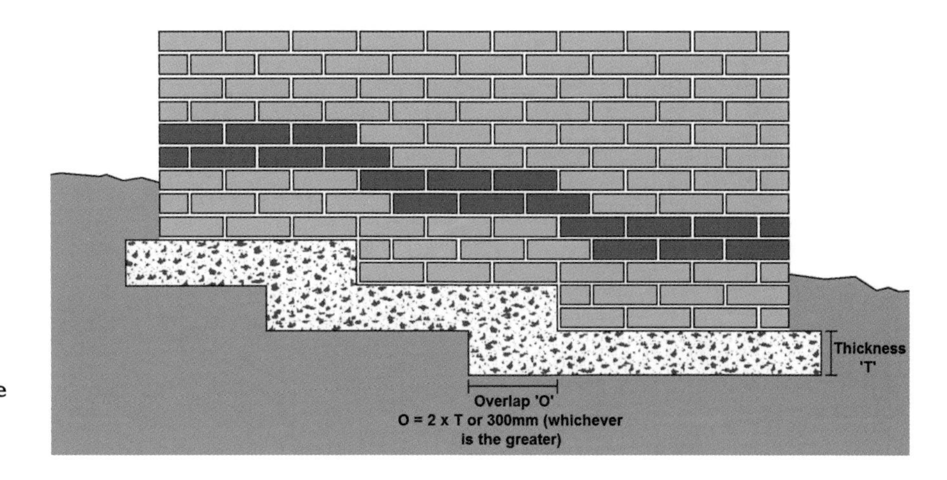

Fig. 41 Concrete stepped strip foundation.

Deep-Strip or 'Trench-Fill' Foundation

A deep-strip foundation, also sometimes referred to as a 'trench-fill' foundation, is particularly useful in sub-soils such as clay, which can move quite violently with changes in groundwater content. A foundation trench is excavated to the width and depth needed and then all of the trench is filled with concrete to a level that stops within two brick courses of finished ground level (*see* Fig 42). The depth ('D') is as for a traditional strip foundation, however, the width ('W') can be reduced to 450mm as there is no longer any requirement for working space within the trench itself.

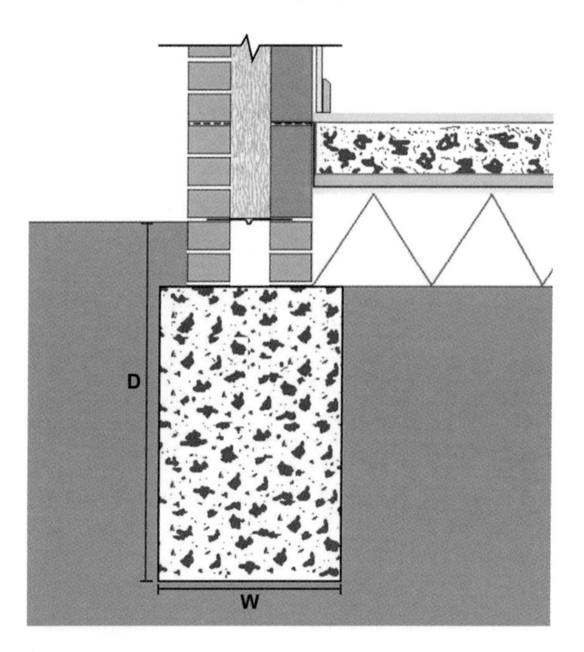

Fig. 42 Concrete deep-strip or 'trench-fill' foundation.

Deep-strip foundations require more concrete but have certain advantages over traditional strip foundations. Excavation is faster as less material is dug out (although most of the spoil has to be disposed of elsewhere, since little back-filling is required). Also, time and materials and skilled labour are saved by not having to construct walling below ground level.

RAFT FOUNDATION

On ground with low load-bearing capacity it is not economical to excavate for traditional strip foundations down to significant depths in order to locate a sufficiently strong sub-soil. Raft foundations offer an alternative and also form the structure for the ground floor. A concrete bed is constructed that is at least equal to the base area of the building. The raft 'floats' on the sub-soil, rather like a raft on water. This method can be used for light buildings or where the top 600mm of sub-soil overlies a substrata of poorer quality. Raft foundations can be varied by thickening sections, particularly the edges, where external and/or structural walls are to be constructed.

There are a number of advantages to raft foundations: they involve relatively little excavation work, they are less expensive to construct than traditional strip and deep-strip foundations, and they are more quickly and easily placed than other foundation types. However, they are generally restricted to lightly loaded buildings or structures.

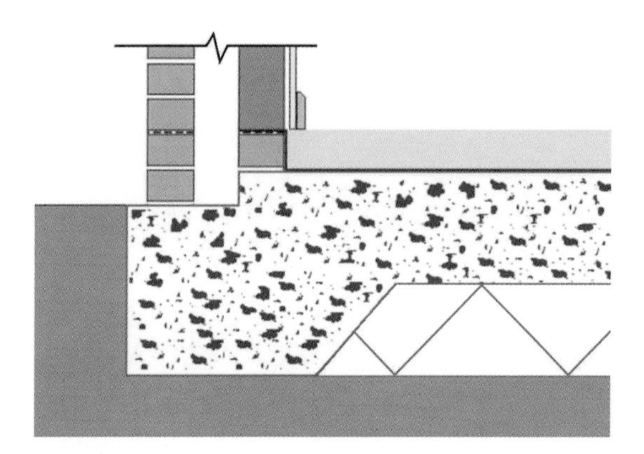

Fig. 43 Concrete raft foundation.

SUB-BASES FOR CONCRETE

Where concrete is placed closer to finished ground level – for example, for paths, patios and bases – it is vital that the soft top-soil, with no bearing capacity, is excavated and removed and that excavation is continued to a depth where sub-soil of a reasonable bearing capacity can be found. The subsequent reduced level is then brought back up using layers of well-compacted hardcore, thus saving on concrete, but also providing a key load-bearing and distribution layer between the concrete and the sub-soil. Hardcore can be clean, crushed rubble or Type 1 MOT stone and, if prop-

Fig. 44 Type I MOT stone.

erly placed, prevents settlement of the concrete and distributes loads through to the sub-soil beneath. Failure to excavate down to a satisfactory reduced level and then provide a hardcore sub-base will undermine the strength of the concrete, resulting in cracks and failure when subjected to significant loading.

The most commonly used material for constructing a sub-base is graded MOT Type I stone. Its correct technical description is 'Granular sub-base material to Type I of the Department of Transport Specification for Highway Works (Clause 803)'. In theory, Type I MOT stone is meant to have a maximum particle size of 37.5mm, but typically it is a crushed rock, usually granite or limestone, with a granular structure of 40mm chunks down to dust.

Fig. 45 'Whacker plate'.

Sub-base material should be laid and compacted in layers, with each layer being thoroughly compacted before the next layer is placed, until the desired overall thickness is achieved. The rule of thumb is that the thickness of each individual layer should be no less than twice the thickness of the largest grain size. On this basis, Type I MOT stone should be compacted in layers no less than 75mm thick (in fact, 100mm layers are typical). The idea of this is that 'point loads' within the sub-base are avoided because no individual stone can be in direct contact with both upper and lower surfaces at the same time and is, instead, 'cushioned' by smaller particles above and/or below. This approach ensures that the layer evenly spreads the loads placed on it and any 'rocking points' within it are eliminated. Sub-base materials using particles of a larger size will, of course, require compacting in thicker layers.

Ideally, when forming a sub-base, mechanical compaction in the form of a 'whacker plate' or roller should be employed in order to ensure maximum compaction of each hardcore layer. This will result in an extremely hard surface that is ready to receive the concrete on top.

Recommended minimum thicknesses of sub-bases for different applications are shown in the following table. These values assume that excavation down to a satisfactory load-bearing sub-soil has been achieved. A sub-base thicker than the minimum stated may be required where a deeper-excavated reduced level needs to be brought back up, in the event that a satisfactory sub-soil is located at a much lower depth. The minimum thicknesses indicated are, therefore, not to be taken as indicating the maximum amount to be excavated beneath the underside of the proposed layer of concrete, as a sub-base cannot be placed on just any type of sub-soil!

Hardcore Sub-Base Thicknesses

Application	*Minimum thickness*
Patios, garden paths etc.	75–100mm
Driveways, public footpaths etc.	100–150mm
Heavy uses	150–225mm
Highways	150mm +

Buried Services

Before excavating for any foundations or bases, great care must be taken to locate any buried services such as drains, electric cables, gas pipes and water pipes, and so on, that may be affected by the excavation. Some services may need to be protected during building work or even permanently diverted.

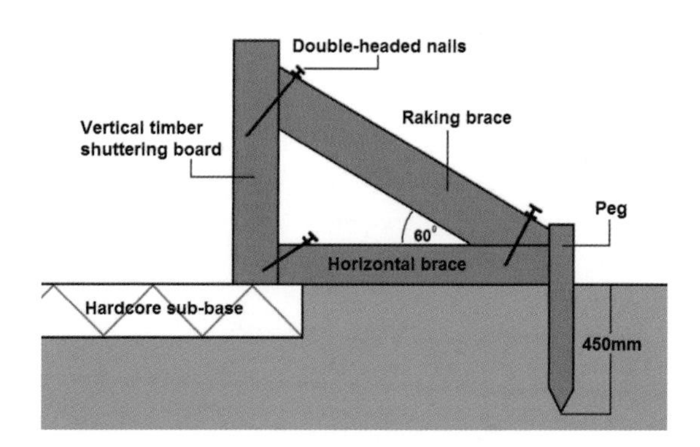

Fig. 46 Timber shuttering/formwork.

SHUTTERING FORMWORK

When concrete is placed with its upper surface below finished ground level, as is the case with foundations for walls, the sides of the excavation itself provide the necessary containment for the concrete, define the plan shape of the concrete and act as shuttering or formwork. The levelling of the top surface of the concrete is simply achieved by positioning timber pegs in the bottom of the excavation, projecting out by an amount equal to the thickness required for the concrete. The pegs are spaced at no more than 1500mm centres along the centre line of the foundation and are levelled with a spirit level and straight-edge or a Cowley level (*see* Chapter 6). When the concrete is poured it can then be levelled to the top of the pegs.

Whilst this chapter does not concern itself with the details of excavation methods, it is worth noting that the process of positioning and levelling pegs to provide both the level and the thickness of the concrete will also give an indication as to whether the bottom of the excavation has been dug out level. If there are any high spots, further grading out or additional excavation may be necessary, otherwise the effective thickness of the concrete will be reduced. Dips or hollows will obviously fill with concrete but excessive low spots should be minimized during the excavation process as additional concrete is an expensive way of compensating for excavating too deeply in places.

When concrete is placed with the upper surface above ground level, extra temporary provision has to be made to contain and form the shape of the concrete as it hardens. The easiest and quickest solution takes the form of shuttering or formwork. For one-off concreting projects such as a shed base or concrete path, this is made from timber; it is bespoke for each project and is not necessarily intended to be reused.

A typical timber formwork arrangement (*see* Fig 46) is held together with double-headed nails so that it can be dismantled once the concrete has hardened, although the use of screws is probably more common. Angled bracing timbers are used, designed to resist the lateral pressure of the wet concrete and to prevent the formwork distorting or bulging outwards. Where the concrete being cast is comparatively thin (around 100mm) then the formwork and its bracing need not be quite so elaborate.

Fig 47 shows the casting of a shed base with the formwork constructed from 100mm × 20mm lengths of timber. In this case, the timbers have

Fig. 47 Simple formwork for a small concrete base.

been screwed together at the outside corners. Additional support has been provided by 20mm × 30mm section vertical timber pegs positioned at the corners and along the sides, with screws inserted through the pegs into the formwork. Additional support to resist deformation during placing of the concrete has been provided by the use of dry bricks.

All formwork is constructed with its upper edge at the same height as the top surface of the finished concrete and is positioned so that it is level all the way round. This assumes of course that the finished concrete is required to be level; falls to the top surface can be easily created by constructing the formwork accordingly. This means that the wet concrete can be levelled as it is placed and compacted with a tamping board. Where necessary, the tamping board can be 'shuffled' from side to side as it is drawn forward along the top of the formwork in order to grade off any excess wet concrete and to obtain a flat surface. Generally, the timber straight-edge used as a tamping board must be long enough to extend 100mm beyond both sides of the formwork in order to be effective. For more on the placing, compacting and finishing of concrete, see Chapter 2.

EXPANSION AND ISOLATION JOINTS

External finished concrete expands and contracts upon wetting and subsequent drying out. This is most significant where large areas of concrete are concerned and in areas where concrete is bounded by other structures such as walls or buildings. If provision is not made for such movement, pressure cracks can appear. These can be deep enough to penetrate through the entire thickness of the concrete, resulting in structural failure.

Movement can be accommodated by the inclusion of expansion joints at 12-metre intervals, sealed with a material that expands and contracts with the concrete but is waterproof, durable and will not allow dirt and grit, and so on, to get into the joint. Similar provision is made where a concrete slab butts up to an existing wall or structure; this type of joint tends to be referred to as an 'isolation joint'.

Fig. 48 Roll of polyethylene expansion joint foam.

The materials for expansion and isolation joints vary, but the most common in a domestic setting are rolls of polyethylene foam. This comes in various widths to suit different concrete slab thicknesses and is positioned against inside faces of the formwork before the concrete is placed.

Longer-lasting expansion joints for more critical applications can be made with various gun-grade or pourable Polyurethane- or poly-sulphide-type compounds. These have a high modulus of elasticity, so they can stretch and shrink more than normal sealers. The packaging is often labelled with 'Hi-Mod' or similar. A 12mm open joint is formed in the concrete by way of a length of softwood or polystyrene foam, which is then removed when the concrete has set. The joint is then filled with the sealer compound as per the manufacturer's instructions; care must be taken not to smear or stain the concrete surfaces as the sealant is very sticky. During installation, it is advisable to protect the surfaces of the concrete on either side of the joint by masking it off with tape or pieces of timber. Once complete, the treated joints are covered with a board until the compound has cured, to prevent it being affected by foot traffic.

CHAPTER 6

Basic Setting-Out and Levelling

KEY POINTS ABOUT SETTING-OUT

Setting-out is the most important process that is undertaken prior to construction work starting. It involves establishing, among other things, the position of the structure on the site, the relative position of walls and all associated levels. The accuracy of horizontal and vertical dimensions is critical if work is to be completed correctly and to acceptable standards.

Mistakes made at any stage during the construction process can be costly to rectify but mistakes made at the setting-out stage can be particularly disastrous. A wrong measurement or wrong angle or being out of square at the setting-out stage can result in large sections of work having to be demolished and rebuilt at great expense.

The basic elements of the setting-out process involve setting-out straight lines in relation to base lines (for example, an existing structure); setting-out 90-degree angles; and methods of transferring levels. The explanations given here are based on a small rectangular building; the basic principles involved can be applied equally to a simpler project such as a straight, free-standing wall, or to a more complex project such as a large building of more complicated shape and layout.

TOOLS, EQUIPMENT AND MATERIALS FOR SETTING-OUT

When setting-out a small building, assuming machine excavation methods for the foundations, the following tools will be needed:

- builder's square
- claw hammer;
- 'long-arm' square;
- lump hammer;
- pencil;
- tenon saw;
- spirit level;
- straight-edge;
- tape measures (5m and 30m; tapes made of fabric are best avoided as they will stretch and give a false measurement!).

Equipment will include a Cowley Automatic Level, a tripod, a target and staff. Materials will include drawings containing relevant dimensions, a nylon ranging line, round-headed 50mm clout nails (galvanized), timber for profile boards and timber for wooden pegs.

THE 'BUILDING LINE'

Setting-out of buildings or structures on site is done in relation to the Building Line. In its strictest sense, this is an unseen line historically allocated to every building plot, beyond which the front face of a building on that plot must not project. It is established by the Local Authority responsible for the geographical area in which the building is located. It is usually set back at a distance from adjacent highways and access roads, so that these might be widened if future development in the locality requires it. Buildings may be placed with their front elevation on the Building Line, or at any distance behind it, with the position having been agreed with the Local Authority at the planning stage. Accordingly, setting-out for the original building will have been carried out

Pegs and Ranging Lines

Setting-out on site makes use of wooden pegs driven into the ground at the corners of the proposed building/wall and nylon ranging lines strung between the pegs to identify the wall lines. Nylon ranging lines are usually orange for the purposes of visibility. Pegs are made from timber measuring 450mm × 50mm × 50mm, with one end sharpened to a point on all four sides so that it can be driven vertically into the ground using a lump hammer.

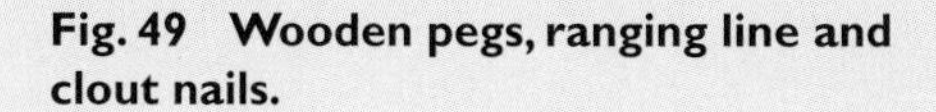

Fig. 49 Wooden pegs, ranging line and clout nails.

with the Building Line as a basis or 'base line' from which to work.

Any new or proposed masonry structure – from boundary walls, paths, retaining walls and concrete bases to house extensions, separate garages, and so on – needs a base line or reference point of some kind from which it can be set-out and in relation to which it can be positioned. This base line or reference point might be an existing boundary line, the edge of a pavement, the wall of an existing building or a wall. The new structure will need to be parallel to this or set-out at an angle from it. Setting-out on plan and establishing dimensions and angles must always be done from a convenient base line!

Fig. 50 Simplified site plan showing proposed new garage.

SETTING-OUT A RECTANGULAR BUILDING

For the purposes of illustration, a project requires the setting-out on site of a rectangular garage in the position shown on the simplified 'site plan' (*see* Fig 50).

A site plan is used to show the position of the proposed building on the site and in relation to any existing structures. It also shows the site boundary, together with information on proposed road, drainage, and service layouts, and other site information such as levels where relevant. The site plan will provide sufficient information, in the form of relative dimensions, to allow the position of the building on the drawing to be transposed on to the actual site.

Selecting a Base Line

The first setting-out task is to select a base line from which all the setting-out for the garage will be derived and established. From the plan (*see* Fig 50), the garage is 5m long × 3m wide. It is parallel to the existing building on the site and is set 1m to the left of it. The front of the new garage is to be set back from the front of the existing building by a

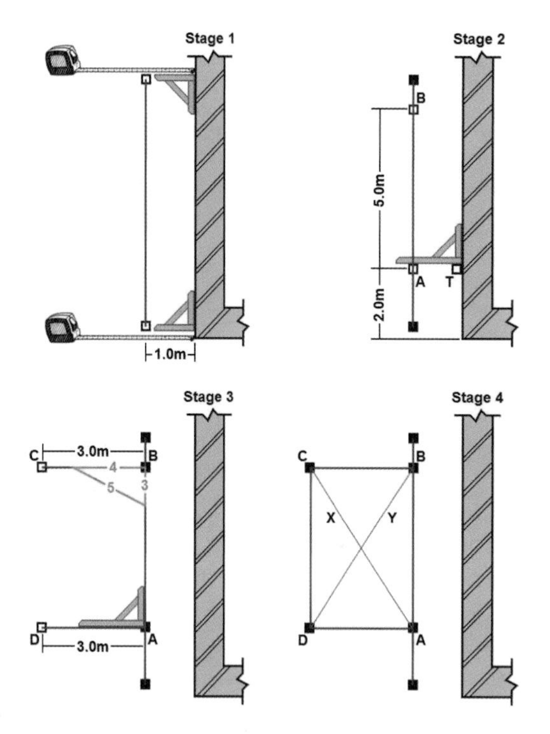

Fig. 51 The four separate stages of setting-out a small garage. Note that this is not drawn to any scale and is merely designed to be a simple illustration of good practice when setting-out.

distance of 2m. The simplest and most convenient base line will be the left-hand side wall of the existing building.

Setting-Out: Stage 1

The first task is to establish the general line of the long side of the garage, parallel to and 1m away from the existing building. As shown in Fig 51, two pegs are positioned 1m (to the centre of each peg) from the side wall of the existing building – these first two pegs are only to establish the side wall line of the garage and need to be positioned away from the corners of the proposed new building.

When setting-out, all measurements are to the centre point of the peg in order to ensure the greatest accuracy at the point of intersection between ranging lines. After positioning and levelling the pegs, a clout nail is hammered into the top/centre of each peg, projecting from the top of the peg by approximately 20mm, to allow the ranging line to be attached.

When measuring the distance from the wall, a builder's square should be used to ensure that the measuring tape is being held at 90 degrees on plan to the wall. Failure to do this will result in a mismeasurement and the pegs will end up being too close to the wall. For the same reason, the tape must also be held horizontally. All setting-out must be carried out in a horizontal plane if correct dimensions are to be achieved. Accordingly, when a string-line is attached to the nail at the top/centre of every peg, it must also be horizontal, so the tops of all the setting-out pegs must be at the same level before hammering in the nail. Methods of transferring levels and ensuring that the tops of pegs are level as shown later in this chapter.

Having correctly positioned and levelled the pegs, hammer a clout nail into the top/centre of both pegs and attach a ranging line. As a secondary check, a line-level can be attached to the ranging line to ensure it is horizontal. Ranging lines must be kept tight at all times to ensure straightness and accuracy.

Setting-Out: Stage 2

Set a temporary peg ('T') a distance of 2m from the front of the existing building to the centre of the peg (see Fig 51), then, using a 'long-arm' builder's square, transfer the position of temporary peg 'T' to the ranging line and insert a peg ('A') under the ranging line. This is the first corner peg of the new garage. Temporary peg 'T' can now be removed.

From peg 'A', measure a distance of 5m and position peg 'B' (which is the second corner peg of the

Fig. 52 Line-level attached to the ranging line.

new garage) level with peg 'A'. Line 'A'–'B' represents the right-hand side wall of the new garage. Insert a clout nail into the top/centre of both pegs 'A' and 'B'.

Setting-Out: Stage 3

The next stage is to establish the remaining two corner pegs by measuring 3m from pegs 'A' and 'B' whilst at the same time squaring from wall line 'A'–'B' so as to be able to insert pegs 'C' and 'D' in their correct positions. The simplest and quickest method of squaring the short wall lines is to use a long-arm square, but this is not the most accurate method. The more scientific and accurate alternative is to make use of the '3:4:5 method' (*see* Fig 51).

Ensure that the tops of pegs 'C' and 'D' are level with pegs 'A' and 'B', then hammer a clout nail into the top/centre of both pegs 'C' and 'D' and attach ranging lines to complete the front and back wall lines.

Setting-Out: Stage 4

Attach a ranging line between the nails in the top of pegs 'C' and 'D' to establish the last wall line. Finally check all the dimensions and then check for square i.e. that all the corners are 90 degrees, by measuring the diagonals, 'X' and ''Y'. If both diagonal measurements are equal then the setting-out is correct. If they differ to any extent, the following sequence should be followed.

1. Re-check all wall dimensions and correct any small errors by repositioning the pegs. Assuming that the original base line has been set-out correctly 1m away from the existing building, note that pegs 'A' and 'B' must only be adjusted along the base line and must not be moved off it. Pegs 'C' and 'D' can be repositioned in any direction.
2. If any errors have been corrected in the first step, re-check the diagonals.
3. If the diagonals are still not equal, pegs 'C' and 'D' can be moved the same amount in one direction or the other, ensuring that ranging line 'C'–'D' remains parallel with line 'A'–'B' at all times. If 'Y' is the shortest diagonal, move pegs 'C' and 'D' forward towards the front of the site until the diagonal dimensions are equal.

   If 'Y' is the longest diagonal, move pegs 'C' and 'D' backward towards the rear of the site.

SETTING-OUT A RIGHT-ANGLE USING THE 3:4:5 METHOD

The majority of buildings and structures are based around straight lines, squares and rectangles, meaning that most corners and junctions will be at right angles. The ability to set-out a right-angle correctly is, therefore, of fundamental importance since a rectangle or square, without care and attention, can easily become a rhombus shape.

A quick method of establishing a right-angle is to use a 'long-arm' version of a wooden builder's square. As it is made from three separate pieces of timber, the accuracy of a timber square is only as good as the quality of its 'manufacture'. Moreover, with use and exposure to weather, a timber builder's square can warp. It may also come loose at the joints, which will further compromise accuracy.

The 3:4:5 method is a reliable alternative for setting-out and checking right-angles, enabling the front and back walls of the building to be set-out from side wall line 'A'–'B'. This will always give a perfect and accurate right-angle. The 3:4:5 method is based on Pythagoras' Theorum, which states that for every right-angled triangle the square of the hypotenuse (longest side) is equal to the sum of the squares of the other two sides. In other words, $A^2 + B^2 = C^2$ (*see* Fig 53). A right-angled triangle with short sides of 3 units and 4 units will always have a longest side (hypotenuse) of 5 units. This simple ratio of 3:4:5 can be applied for the setting-out of any right-angle – all that is required is that the measurements used are in a ratio of 3:4:5, for example, a triangle with sides 600mm by 800mm by 1000mm, or (3 × 200mm) by (4 × 200mm) by (5 × 200mm).

For a practical application of this method for the front wall line of the garage, *see* Fig 54. The two string-lines are marked at 3 units and 4 units respectively by measuring out from corner peg 'A'. The distance on the diagonal (hypotenuse) between marked points 'X' and 'Y' should measure 5 units. If not, peg 'D' and the line attached to it should be moved until there is an exact measurement of 5 units between 'X' and 'Y', which confirms that the angle at corner peg 'A' (in other words,

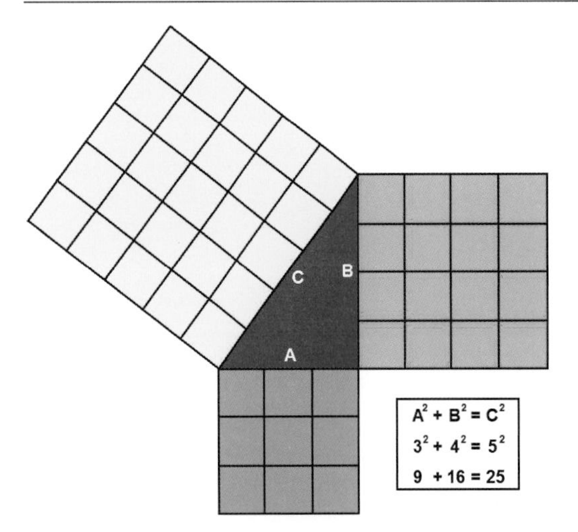

Fig. 53 Pythagoras' Theorem as the basis for the 3:4:5 method of setting-out a right-angle.

the angle formed between lines 'A'–'B' and 'A'–'D') is 90 degrees.

PROFILE BOARDS

Setting-out a building or a wall with ranging lines attached to pegs driven into the ground at the corners is all well and good until it comes to the time to excavate for the foundations. Corner pegs will be in the way. The pegs will have to be removed to allow excavation to take place, which means that all the original setting-out will be lost.

To avoid having to go through the whole setting-out process again in order to establish the position of the walls on the new foundation, it is necessary to find a way of holding ranging lines in their correct position above the ground but away from the point of excavation. This is done by transferring the ranging lines to profile boards that are erected away from the confines of the outer wall lines of the building or structure, in a position where they will not be disturbed by the excavation. This will typically be around 500mmm away from the point where the foundation trench is to be excavated, but it will need to be further (sometimes up to several metres) in case of mechanical excavation, to allow for machinery access.

The profile itself comprises a horizontal board (approximately 75mm × 30mm × 850mm), face-fixed through two supporting pegs (*see* Fig 55). When profiles have been erected all the way round the building, the ranging lines can be detached from the nails in the top of the corner pegs and then extended back to the profile boards, making sure that they precisely retain their original linear position. This is achieved by keeping the ranging line tight and just in contact with the nail in the top of the corner peg (in other words, its original fixing point) as the line is extended back to the profile board. The line is held carefully in place and the top edge of the profile board is marked with a pencil. Where the profile has been marked, a clout nail is inserted or a shallow saw cut made, ready to secure the ranging line.

When all the profiles have been marked and made ready to receive the ranging lines, the original corner pegs can be removed and the ranging lines attached to the profile boards. At this stage, it is wise to re-check the setting-out for both square and accuracy of dimensions; this should be done at least every time the ranging lines are altered, or taken off and re-attached. Pegs and profile boards can remain in place for extended periods of time so it is good practice to check the accuracy of the setting-out at the start of every working day – there is always a possibility that pegs and/or profiles could have been disturbed at some point.

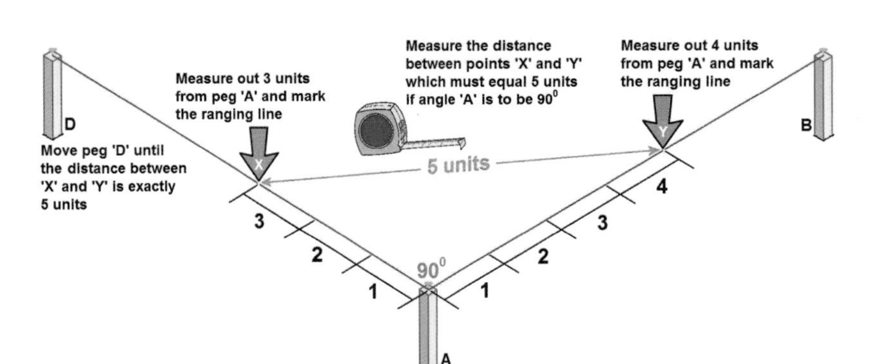

Fig. 54 Practical application of the 3:4:5 method to set-out a right-angle.

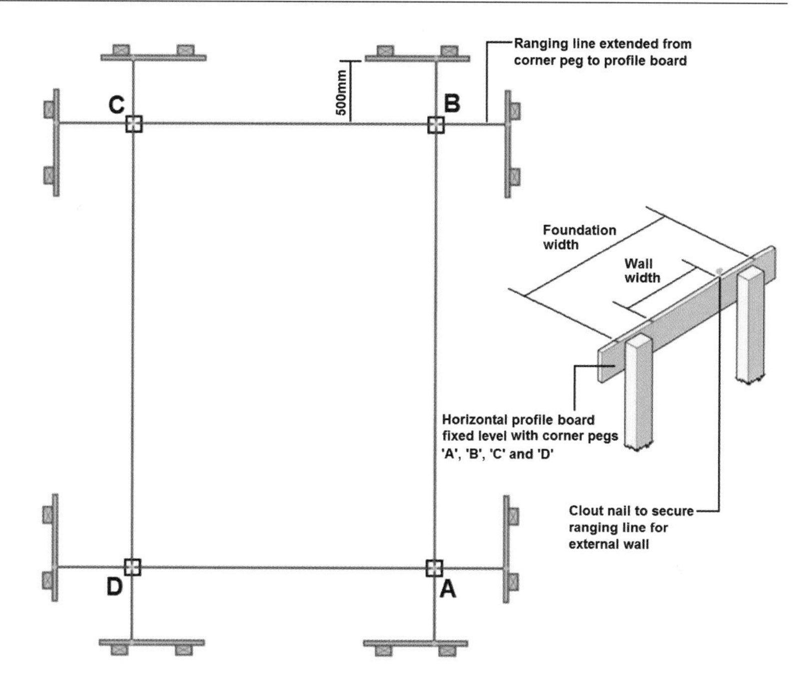

Fig. 55 Setting-out using profile boards.

In the example (*see* Fig 55), the external wall lines of the garage are shown on plan extended to timber profiles.

Profiles can contain additional information, such as the width of the foundation, where extra ranging lines can be attached so that excavation lines can be marked out on the ground. This can be simply achieved by spreading a thin line of lime immediately below the ranging line, or by using spray paint specifically made for the purpose. Some excavator drivers prefer the centre line of the foundation trench to be marked in addition or instead, as it allows them accurately and easily to centre the excavator's bucket.

Once excavation has been carried out and the concrete foundation placed, the ranging lines can be re-erected at the external wall-line positions, and then checked for accuracy. After this, the wall lines can be transferred down to foundation level and the brickwork may be set-out ready for construction.

Levelling Profile Boards

The need to level corner pegs, and the various reasons for the process, also applies to profile boards. The profile board must be horizontal and all profile boards must be level with one another. Ideally, and for accuracy of extending and re-establishing the ranging lines, the profile boards should be set at the same level as the original corner pegs. As the distance between the original corner pegs and the profile board is only 500mm, the level is easily transferred using a spirit level.

SITE DATUM

All vertical dimensions and levels on new walling and buildings are derived from one fixed point on the site. This is referred to as the 'site datum' and is generally established on site prior to any construction work starting. Site datum may also be referred to as a 'temporary bench mark' (TBM).

Site datum level is most commonly established on site at the same level as the horizontal damp proof course (DPC) of the proposed building or wall. For a completely isolated structure, datum can simply be established at a convenient point near to the proposed structure by inserting a wooden peg with its top at 150mm above finished ground level; 150mm is the minimum DPC height above finished ground level for new buildings as required by the Building Regulations. For a boundary or garden

Sloping Sites

For fairly obvious reasons, on sloping sites, datum should be positioned at the higher end of the site so that levels are transferred downhill and not up. When setting-out and transferring levels on a sloping site it is recommended that the setting-out should begin with short pegs at the high point of the site, with longer pegs used as setting-out progresses down the slope. When transferring levels uphill from a low starting point there is a risk of running into the ground before reaching the last peg, so setting-out should always be done downhill!

It is worth remembering that the true extent of a slope can be very deceptive when looking with the naked eye; in fact, a site often slopes much more than it first appears.

wall, despite there being no statutory requirement, there is little reason to deviate from this minimum height dimension.

For new walls or buildings that are adjacent or near to an existing structure, it is common to derive datum for the new brickwork from DPC level of the existing building. A peg is simply inserted at DPC level adjacent to the existing building, then the datum level is transferred to the desired location using either the spirit level and straight-edge method or a Cowley level.

The 'datum peg' should be in a position on site where it can be seen and easily accessed, but where it will not be vulnerable to any damage from machinery or falling materials or passers-by. Maintaining the accuracy of the site datum is of fundamental importance since all vertical dimensions and levels for the new walling are derived from it and a site datum that alters during the course of the work can have disastrous consequences.

The site datum peg should be driven into the ground at its required location and then, ideally, surrounded by concrete (*see* Fig 56). The bigger the project, the longer the datum will be required, so consideration should be given to providing it with additional protection in the form of a triangular timber framework, as shown.

Once datum has been established it must be transferred to both ends of the wall being built, or all the corners for a new building, again, using either

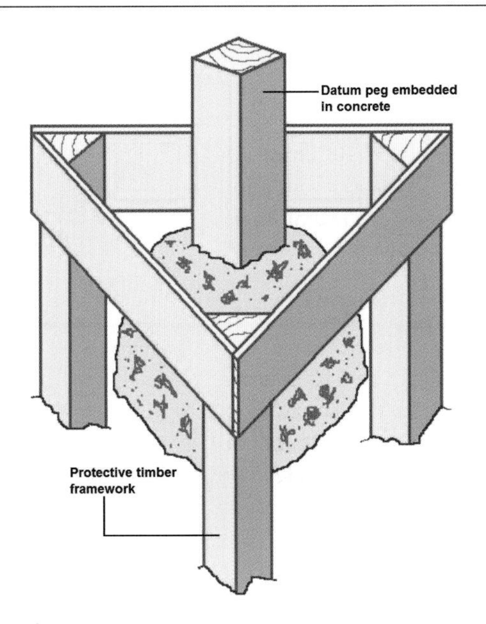

Fig. 56 Site datum peg with protective timber framework and concrete.

a spirit level and straight-edge, or a Cowley level. Concrete foundations cannot be relied upon to be flat or level, so datum is needed at every corner of a building or end-point of a wall, so that the bricklayer, when constructing the corners, can check the gauge of the brickwork from the top of the foundation concrete. In this way, any adjustments to gauge ('picking up' or 'grinding down') can be made below ground level, thus ensuring that all the brickwork will be the same level by the time DPC level is reached. For more on the method of checking gauge from the top of foundation concrete, *see* Chapter 9.

In order to avoid the need for making adjustment to gauge below ground, the datum peg may be used as a basis to calculate how deep to excavate the foundation trench so that, when the concrete has been placed, the vertical distance between the top of the foundation and datum level works to the gauge of the brickwork without the need to adjust the thickness of bed joints (in other words, in multiples of 75mm).

As an example, a simple strip foundation would require a minimum thickness of 150mm and assume a ground cover between the top of the concrete to finished ground level of 1000mm. Assuming that datum has been established at DPC level of 150mm above finished ground level, the overall distance from datum to the bottom of the foundation trench would be 1300mm (calculated as 150mm + 1000mm + 150mm), with the top of the finished foundation being 150mm higher at a depth of 1150mm below datum. This total of 1150mm divided by a brickwork gauge of 75mm works out to 15.33 courses of brickwork from the top of the foundation to datum level. Clearly, this will not work to gauge – 15 courses would finish short and 16 courses would finish too high. Given that the shortfall is only 0.33 of a course (approximately 25mm), ordinarily the bricklayer would choose to adjust gauge by 'picking up' the bed joints as the brickwork is built out of the ground. (For more on basic bricklaying skills, *see* Chapter 9.)

There is an alternative to picking up the bed joints and that is to round up to the next full course. This would mean making 16 courses from foundation level to datum level but excavating the foundation trench a little deeper to maintain a gauge that requires no adjustment. Rounding down to 15 courses is also feasible, but this will mean raising the level of the foundation, with the possible risk of compromising any minimum foundation depth requirements. On the basis of rounding up to 16 courses, the depth of the foundation trench will need to be excavated to a depth below datum level of 1350mm, that is to say, the sum of 16 courses × 75mm + 150mm (for the concrete thickness).

Clearly there is a judgement to be made by the bricklayer, based on the following options: the bed joints may be picked up to gain 25mm, or a 75mm gauge may be maintained by going to the trouble of setting-out and excavating an additional depth of 50mm along the whole of the foundation trench, and laying an extra course of bricks. Obviously, the latter would take additional time, the extra excavated soil would need carting away and there would be an increase in the quantity and cost of bricks and mortar. In most cases then, the convenience of not having to adjust the thickness of bed joints comes at a high price that is never really likely to be viable. Rounding down to 15 courses avoids all the extra excavation and additional bricks and mortar, but still requires time to set-out the revised depth accurately. Most bricklayers would probably still opt to pick up bed joints to gain the 25mm required.

TRANSFERRING LEVELS

Setting-out must be carried out in a horizontal plane. The tops of the pegs for setting-out must be level so that ranging lines can be attached horizontally and accurate plan measurements may be achieved.

There are two simple methods of transferring levels from one point to another: using a spirit level and straight-edge, or a Cowley Automatic Level. The two methods are not limited to levelling the tops of setting-out pegs. Both can be equally applied to any setting-out situation involving the principle of transferring levels between two points, such as transferring the site datum level to the corners of the building or levelling the tops of pegs in the bottom of an excavation to set the thickness of foundation concrete.

Method 1: Spirit Level and Straight-Edge

Transferring levels over a long distance between one point and another, such as one corner of a proposed building to another, cannot be achieved with a spirit level on its own, as the span between the two corner pegs will be too great. Even using a 3-metre straight-edge to extend the effective length of the level is unlikely in the vast majority of cases to prove effective in spanning between one corner peg and the next.

Any length of sturdy timber can be used as a straight-edge, provided its sides are parallel and the timber is, of course, straight. This may seem obvious but timber, even when it has been accurately planed all round, has a tendency to warp, especially when it is exposed to the conditions on a building site. As a result, it is better to use a straight-edge that has been purposely machined and shaped out of good-quality, seasoned timber. Better still, a straight-edge made out of aluminum box-section will always remain straight and true.

Transferring a level from one peg to another peg some distance away must be completed in stages:

1. Drive a temporary intermediate peg into the ground along the path of the straight

line between the first and last peg. It should be placed at a distance from the first peg that is a little less than the length of the straight-edge.
2. Using the spirit level and straight-edge together, check the peg as it is hammered into the ground until it is level with the first peg.
3. Depending on the distance involved, repeat the process with additional temporary intermediate pegs until the position of the final corner peg is reached. At this point, all the temporary intermediate pegs can be removed

When transferring levels in this way, the accuracy of the spirit level should be checked, and adjusted if necessary, before use. In any case, it is regarded as good practice to reverse the spirit level and straight-edge every time they are moved on to a new peg, turning them end-to-end through 180 degrees, in order to compensate for any slight inaccuracy. If there is any inaccuracy, simply moving them along without turning them will exacerbate the error. Reversing them at every peg may create an 'up and down' effect but by the time the last peg is reached, the first and last pegs will be level with each other, provided the number of temporary intermediate pegs is an odd number making the number of stages an even number. This compensatory measure will not work as effectively with an even number of pegs and an odd number of stages.

Method 2: Cowley Automatic Level

When transferring levels over very long distances, the use of a spirit level and straight-edge can be very time-consuming and prone to error.

The Cowley Automatic Level (or just 'Cowley' as it is commonly known) is a simple and, by modern standards, somewhat old-fashioned type of levelling instrument. Despite not having the same sort of complexity as modern optical and laser levels, it is readily available to be hired for simple and/or one-off levelling tasks and remains a very effective levelling tool. Its accuracy is limited to around 30m in one direction but it can be rotated through 360 degrees, making it effectively capable of transferring levels between pegs that are 60m apart, when it is placed in the middle. This range should be more than adequate for most applications and projects.

The equipment consists of three main components (see Fig 58):

1. A metal-case head unit that houses a sensitive, balanced system of prisms and mirrors, some of which are controlled by pendulums, which ensures that the sighting is always along the same line.
2. A tripod with a metal spike on top on to which the Cowley level is placed.
3. A graduated staff with an adjustable target that slides up and down the staff. The target is usually 450mm long and 50mm wide and is painted in a fluorescent orange, red or yellow paint so as to be easily visible.

The Cowley level is set up on the tripod and adjustments are made to the tripod legs to level

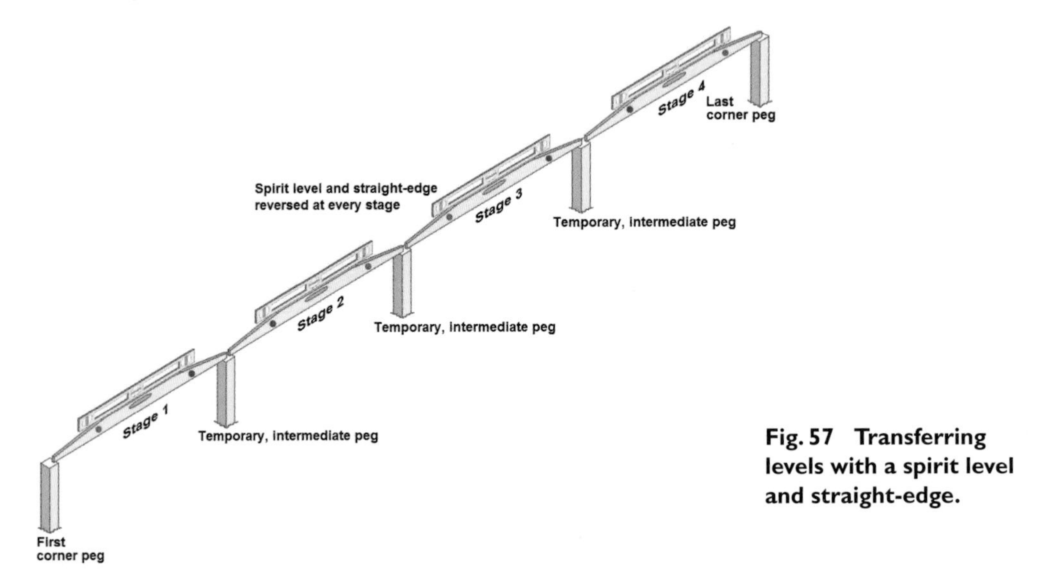

Fig. 57 Transferring levels with a spirit level and straight-edge.

Fig. 58 Cowley Automatic Level, tripod, staff and target.

up the instrument by eye – it does not have to be perfectly aligned in order to do its job. Fig 59 illustrates the general operating procedure for a Cowley level, together with the potential results that will be seen through the eyepiece when looking at the target.

Note: the action of placing the head unit on the spike releases an internal locking mechanism that allows the mirrors to move freely. This is why the level must never be carried around when attached to the tripod as this can damage the internal mirror and prism system.

Once the tripod legs have been adjusted, the staff can be positioned on the site datum peg, or any peg whose level is to be transferred to other points on the site. Rotate the Cowley level until the target is in view through the eyepiece – it will appear as a spilt image of its two halves side by side. Slide the target up or down the staff until the two halves of the target coincide exactly with one another. Lock off the target with the screw at the back and then check that it is correct and has not moved whilst being locked off. The Cowley level and the target are now in the same horizontal plane.

The instrument can now be rotated through 360 degrees to transfer the level to other pegs established at any point within its range. The position of the target on the staff remains fixed and each new peg is driven into the ground to the required depth so that the view through the eyepiece reads level at every new peg.

CHECKING SETTING-OUT

The most important sequence is to set-out, check and re-check before building work commences. It is vital to be absolutely sure that the dimensions are accurate and the location of the structure is correct, and it is not possible to check the accuracy of setting-out too many times! If setting-out work has to be left for an extended period, such as overnight, it should be checked again in the morning to ensure it has not been disturbed. Dimensions and diagonals should be checked every time ranging lines are removed and re-attached.

Successful setting-out cannot be carried out by one person. It requires at least two people but it is preferable to have three, due to the number of processes that have to be brought together at the same time.

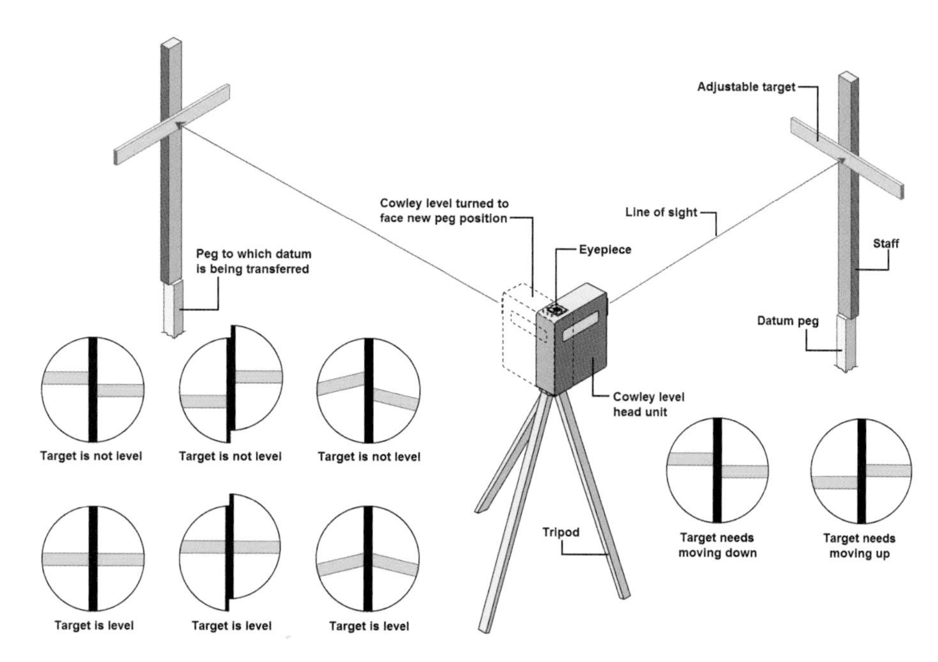

Fig. 59 Cowley level operating procedure for transferring levels.

CHAPTER 7

Bricklaying Tools

TYPES AND QUALITY

A bricklayer's basic tool kit includes tools required for four general types of work: spreading; levelling and measuring; pointing and jointing; and cutting bricks and blocks. The tools covered here are those that a bricklayer should ideally possess in order to cover the basic craft skills; not all can be bought and a few will need to be home-made. There are a number of other tools that are used infrequently, as and when necessary, that would not be regarded as an essential part of a basic bricklaying tool kit. These are not detailed here but are referred to elsewhere at the point at which they are relevant.

Unless tools are being bought for one-off or very infrequent use, there is little advantage in buying cheap ones, since they will be of comparatively low quality, are unlikely to be robust enough to stand up to the rigours of frequent use and represent generally poor value for money. It used to be a general rule of thumb that a good-quality tool would have the maker's name permanently etched, stamped or engraved into it. This still applies today but to a much more limited extent, due to the nature of modern mass production. However, the construction industry is quite conservative and many bricklayers continue to use tools that have been tried and tested and made to a high standard by well-known manufacturers for years. For example, Spear & Jackson is synonymous with the manufacture of quality, solid-forged trowels, as Stabila is with the manufacture of spirit levels.

If tools are going to perform in the way they are designed to, they must be looked after, maintained, sharpened when required and not be misused or abused. Tools should be used only for the purpose for which they are intended, and at the end of the day's work they should be cleaned, dried and properly stored away. Prior to storage, some tools, such as trowels and jointing irons, will benefit from being wiped with a lightly oiled cloth to protect them from rust. Tools will give many years of service and some will last a lifetime but only if they are used and maintained properly. On a building site, the condition of a bricklayer's tool kit is a good indication of his or her quality of work and attitude!

SPREADING TOOLS

Brick Trowel

The first and essential item in a bricklayer's tool kit is the brick trowel (Fig 60), which is used for cutting, rolling and spreading mortar. A good-quality trowel is forged and milled from one solid piece of steel; the kite-shaped blade (1) and tang (5) are one and the tang extends through the full length of the handle (7). Handles are made either

Fig. 60 London Pattern broad brick trowel.

of plastic with a soft-grip rubber outer coating or hardwood and are terminated at the back with a cap (8) that is riveted through into the tang. The cap used to be made of steel but today most are made of a hard rubber. The point at which the tang enters the handle is finished with a metal ferrule (6). The London Pattern trowel (Fig 60) is available in lengths from 225mm up to 350mm, and in two widths, 112mm (narrow) and 138mm (broad). The latter also tends to have a slightly thicker blade and is, therefore, a little heavier. It should never be assumed that a bigger trowel will lead to faster bricklaying; trowels should be selected entirely on what feels comfortable for the user.

All London Pattern blades are ground to be thicker at the heel (4) than at the tip (3), which improves the balance and handling of the trowel. The outer edge of the blade of a brick trowel is tempered, as it needs to be harder in order to withstand being used for tapping bricks to level, and the rough cutting of soft bricks. (It is good practice to avoid the latter and use the correct cutting tools instead.) The tempered 'cutting edge' (2) has more of a curve than the inner edge, which also makes it easy to differentiate between right- and left-handed trowels. Left-handed trowels tend not to be stock items and will need to be specially ordered. The inner edge is generally straight as it is used for trimming off excess mortar and acts as a guide to the alignment of bricks during the laying process.

In recent years, some bricklayers have moved away from the London Pattern trowel in favour of the Philadelphia Pattern trowel, which tends to have a thinner, lighter and more flexible blade. In addition, the back end of the blade is much wider and the blade edges much more curved, which, overall, is intended to make it possible to pick up more mortar. The handle of the Philadelphia Pattern trowel is also noticeably longer, which, allied to the overall lightness of the trowel, is intended to reduce fatigue in the hands, wrist and forearm. In terms of quality, the same solid forging of blade and tang should be sought.

Fig. 61 Philadelphia Pattern brick trowel.

Fig. 62 Stainless steel floating trowel.

The trowel should be cleaned after every use and the blade wiped with an oily rag in order to prevent it from rusting.

Steel Floating Trowel

Floating trowels are used to spread and apply a surface to screeds, renders and concrete. A bricklayer will most commonly be using it for trowelling concrete to provide a surface finish. The better-quality floating trowels are made from stainless steel. Another indicator of quality is the number of rivets used through the underside of the blade to attach the handle – the more rivets (eight to ten is a good number), the better the quality.

Again, the trowel should be cleaned after use and the blade wiped with an oily rag.

Wooden or Polyurethane Float

A float is used in a circular motion to flatten out the surface of concrete prior to applying a surface

Fig. 63 Polyurethane float.

finish with a steel floating trowel. For the most part, floats used to be made of wood and it was not unheard of for bricklayers to manufacture a makeshift float from off-cuts of timber. These days, modern floats tend to be made from polyurethane and are often referred to as 'poly floats'.

Fig. 64 A selection of spirit levels.

LEVELLING AND MEASURING TOOLS

Spirit Level

The ability to obtain a measure of true horizontal (level) and vertical (plumb) surfaces is of pivotal importance to a bricklayer, both when laying bricks and as part of the setting-out process prior to construction.

Horizontal level and vertical plumb are obtained by using a spirit level, which has existed, in one form or another, since the seventeenth century. A spirit level comprises an elongated steel frame, lightweight aluminium box section or, sometimes, hardwood straight-edge, into which are fixed holders or 'vials'. The vials contain short, sealed tubes of glass which have a slight radius to them. The tubes are almost completely filled with spirit but with sufficient air left inside to create a bubble that will sit centrally when the radius is aligned at the top. Most spirit levels will incorporate one longitudinal vial for horizontal levelling and one or two transverse vials, at the end/s, for vertical plumbing.

A spirit level is more than just an air bubble inside a tube of liquid; it is also an invaluable straight-edge for measuring the line, flatness and face plane of brickwork. It will provide a bricklayer with vital clues and information about how bricks are positioned and the extent and direction of any adjustment that may be needed.

Sizes

Spirit levels come in many sizes, from 600mm to 2000mm long, but most bricklayers tend to favour a 900mm or 1200mm level. If it is any bigger it can become rather unwieldy to use, particularly in a confined work area. Often, a spirit level will be used in conjunction with a longer straight-edge, in order to transfer levels across distances that exceed the length of the spirit level.

Shorter levels of around 300mm long are available for small levelling work such as individual bricks within a decorative feature or for plumbing up soldier bricks or bricks laid on edge (for example, for a coping on top of a boundary wall). These are commonly referred to as 'boat levels', as traditional versions resembled the plan shape of a boat.

Checking the Accuracy of a Spirit Level

A spirit level is designed to be robust for use on site but it is still a precision measuring instrument, which should not be abused or misused. It should not be dropped on to a hard surface, and hitting it with a trowel to adjust bricks that are out of level or plumb is bad practice and the sign of a poor bricklayer.

A good-quality spirit level comes with a vial that can be adjusted and all spirit levels should be checked from time to time for accuracy, if the quality of the brickwork on which it is used is to be maintained. The obvious method is to check the device against a surface that is known to be level and/or plumb but often this is not possible or completely reliable. The following method is a good alternative (see Fig 66):

1. Insert two screws in a generally level surface a distance apart just less than the length of the level.
2. Place the level on top of the screws and adjust one of the screws until a level reading is obtained.
3. Turn the level through 180 degrees and note the new reading. If the reading still shows level then no adjustment of the vial is necessary. If, however, the reading shows an error, half the error should be corrected by adjusting the vial and half by adjusting the screws.
4. Turn the level through 180 degrees again. It should now read accurately in both directions.

The same method can be used to check accuracy for plumb, by using two screws fixed into a generally vertical surface.

Fig. 65 Boat level.

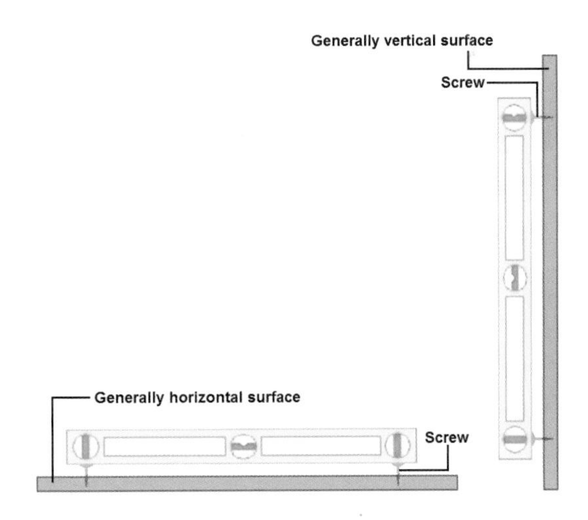

Fig. 66 Checking spirit levels for accuracy.

Tape Measure

Tape measures are used extensively in construction and are available in various sizes from 2 metres up to 30 metres long. They are made from thin, continuous metal or reinforced fabric, about 13mm wide, which enables them to be wound up tightly within a spring-loaded metal or plastic case. Fabric tape measures have a tendency to stretch with age and extensive use, and are not recommended where particularly accurate measurements are required. A tape measure of up to 6 metres is small enough to carry on a belt clip or in the pocket. Longer

Fig. 67 Tape measures.

Fig. 68 Gauge lath.

tape measures tend to be used for long-distance work such as measuring boundaries or setting-out for foundations and wall positions. It is usual for a bricklayer to have a small tape (3m to 5m) for day-to-day use and a longer tape (20m to 30m) for larger setting-out tasks.

Most tape measures have dual measurements marked on them in metric and imperial, with metric being the more commonly used. The case for 2m to 6m tapes is usually of a specified size so that it can be part of the overall measurement when measuring, for example, between the vertical reveals of openings.

Tapes should be kept wound up when not in use and a steel tape should not be left lying around on the ground; if it is trodden on, it is liable to kink or split. Pulling retractable tapes fully out of their casing beyond the last graduation is likely to damage the spring-loaded retraction mechanism. Long tapes, when extended and being moved round a site during setting-out, for example, should be carried by two people and not dragged along the ground, as this could result in damage. The tape must be kept supported and not allowed to kink. If working in muddy or wet conditions, it should be run through a clean, dry cloth as it is retracted. A steel tape should also be lightly oiled at the end of each day, to protect it from rust.

Gauge Lath

Gauge is the vertical measure applied to the height of brickwork in terms of the combined thickness of bricks and the mortar bed joints between courses. Bricklayers will often check gauge by measuring with a tape, in multiples of the thickness of a course of bricks. For example, 65mm brick + 10mm bed joint = 75mm gauge. Typically, and for convenience, bricklayers will check that every fourth course measures 300mm. There are measuring tapes available that are specifically marketed at bricklayers, marked out for the purpose of gauging brickwork.

A more traditional alternative still favoured by many bricklayers, however, is a gauge lath that is marked with 75mm graduations, made from a length of timber 50mm × 20mm in section (or similar). This device makes gauging the height of every course a simple task. The marks on a gauge lath are made with permanent, shallow saw cuts rather than a pencil.

Squares

Bricklayers will often have a small wooden square for the purposes of checking small work – for example, checking that individual bricks in a brick-

Fig. 69 Steel builder's square and small wooden square.

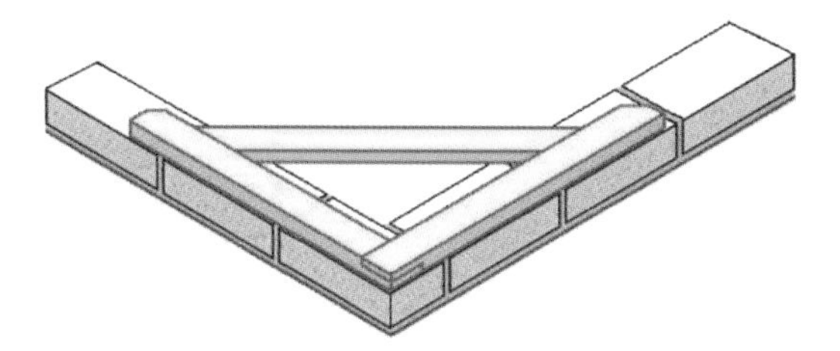

Fig. 70 Wooden builder's square.

on-edge coping are being laid square across the top of the wall at 90 degrees to the front wall line. Such a square, 150mm × 150mm, can be cut in one piece from the corner of a sheet of plywood.

A bricklayer will usually also have a larger square with arms that are around 700mm long, either manufactured in steel (Fig 69) or home-made from wood (Fig 70). The steel version does not need any kind of brace between the two arms, making it easy to use when squaring brickwork on both internal or external corners. A wooden builder's square, being made from three separate sections, needs a brace that ties the two arms together, so it is a little more awkward to use when squaring an external corner. Because it is home-made from three pieces of timber, its accuracy may be questionable. Moreover, with use and exposure to weather, a wooden builder's square can warp and can also come loose at the joints, and this will further compromise accuracy. On this basis, a steel builder's square is generally preferable. 'Long-arm' versions of both steel and wooden squares are used for the setting-out process on site.

POINTING AND JOINTING TOOLS

Pointing Trowel

A pointing trowel is a small version of a brick trowel and is made in the same way, with a blade length of anything from 50mm to 150mm. Its primary function is to apply mortar into joints when pointing and to apply struck joint finishes to mortar joints.

When new, the blade of a pointing trowel has quite a sharp point and this tends to interfere with the ability to apply a struck finish to cross-joints – the point tends to dig in to the joint and ruin it when the bottom of the cross-joint has been reached with the downward stroke of the pointing trowel. For this reason, bricklayers will grind off to a neat curve the top 10mm of the tip of a new pointing trowel.

Fig. 71 Pointing trowel.

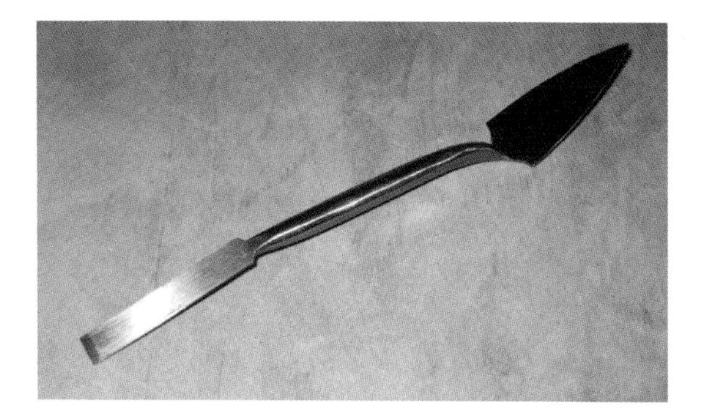

Fig. 72 Small finger trowel.

Finger Trowel

From time to time, bricklayers will encounter pointing and jointing tasks where access is limited and a pointing trowel may not be accommodated, for example, when pointing round a window frame or in a tight internal corner of brickwork. In such cases, a small steel finger trowel is an ideal alternative.

Half-Round or 'Bucket Handle' Jointing Iron or Jointer

The term 'bucket handle' is a throwback to a

time when bricklayers would remove a galvanized handle, with a half-round profile, from a bucket, then straighten it and use it as a jointing tool. Nowadays, there are various bespoke half-round jointers manufactured from steel and available to buy, but the term 'bucket handle' remains in use to describe both the tool and the joint finish.

Despite the availability of manufactured jointers, many bricklayers still prefer to make their own from a length of 13mm-section mild-steel bar, cranked to the shape of a jointer (see Fig 73). The 'handle' is provided with a grip by slipping on a short length of old hosepipe. Using 13mm bar means that the jointer is wider than the mortar joints, so it cannot be pressed too deeply into the joint. One key reason why bricklayers favour a home-made jointer is because manufactured jointers tend to have quite a short blade (around 100mm), so that it is difficult to keep the tool straight and true when jointing long bed joints. This tendency for a short jointer to 'wander' can result in a lack of uniformity in the joint finish. The blade of a home-made jointer can be made to any length required by the bricklayer – typically, it will be around 200mm long.

Fig. 73 Half-round mild-steel jointer.

Fig. 74 Brick hammer.

Fig. 75 Head of a scutch hammer showing the replaceable comb.

CUTTING TOOLS

Brick Hammer

A brick hammer is predominantly used for rough cutting and trimming the excess from cut bricks, although the bladed end, with care, can also be used for trimming the fair-faced edge of a cut if the bolster chisel has not quite done its job.

An alternative to the brick hammer is a 'scutch', which, instead of a bladed end, has a slot into which a replaceable comb or blade is inserted. Its main advantage over a brick hammer is that the main tool never wears out, whereas continued re-sharpening will eventually wear out the brick hammer.

Lump Hammer

The heavy lump or 'club' hammer, which weighs around 2kg, is used in conjunction with a variety of cutting chisels. As with all hammers, the metal head can work loose from the wooden handle with use so it is a good idea, from time to time, to re-secure the head. This is done by soaking the head of the hammer in water and hammering in the cleats that are fixed into the top of the handle.

Fig. 76 Lump hammer.

Sharpening Tools

All chisels and cutting tools must be kept sharp and the use of blunt cutting tools must be avoided. The compensatory over-exertion necessary to cut with blunt tools results in more personal injuries and/or damaged/ wasted bricks than are ever attributed to the use of sharp tools. Percussion tools, such as all types of chisel, will, with the impact of the hammer, start to deform and form a 'mushroom' head of metal burrs at the end of the handle. This can cause serious injury if the chisel, when being struck by the hammer, slips through the hand of the user. From time to time, it is vital that the deformed metal is removed with a grinder or grinding wheel in order to maintain the safety of the tools in use.

Fig. 78 Plugging chisel.

Fig. 77 Bolster chisel.

Bolster Chisel

A bolster chisel is used for fair-faced cutting of bricks and blocks. For this purpose it should ideally have a blade width of 100mm. A bolster with a narrower blade tends to be used more for chopping out toothings and indents in masonry.

Plugging Chisel

A plugging chisel is used for cutting out mortar joints around bricks that are to be removed or for removing old, perished mortar prior to re-pointing.

Cold Chisel

A cold chisel is a heavy-impact tool used when cutting holes, toothings and indents in masonry.

Fig. 79 Cold chisel.

Fig. 80 Narrow bolster chisel with worn 'mushroom' head.

Fig. 81 Bricklayers' lines and pins.

OTHER TOOLS

For all walling that exceeds the length of a spirit level, a string-line must be used for running-in to ensure good and straight alignment of brickwork in between the two ends. The string-line is attached at each end/corner of the wall using one of two basic methods.

Line and Pins

The first method uses a pair of steel pins (*see* Fig 81), approximately 125mm long with sharp, flattened ends, pushed into the joints of the brickwork. The use of pins has a number of disadvantages. The pins physically damage the mortar joints, which must be made good later; clumsy use of pins can disturb the brickwork itself; and it can often prove difficult accurately to align the line with the top arris of the bricks, which is made worse by the habit that pins have of working loose. Most bricklayers avoid using pins unless they have to; the main use of line pins is as something to wrap the string-line around when it is not in use.

Corner Blocks

The second method is the use of 'L'-shaped corner blocks, which can be made from off-cuts of timber or bought as plastic versions in pairs. There is no fixed size for corner blocks but one block would be of sufficient size if it were cut from a block of wood of maximum size 75mm × 50mm × 50mm, or similar. Corner blocks are held in place by the tension of the line and have the distinct advantage of not damaging the mortar joints. However, they can only be used on external corners, and the use of pins is sometimes therefore unavoidable.

Fig. 82 Wooden corner blocks.

CHAPTER 8

Bonding of Brickwork and Blockwork

BONDING OF BRICKWORK

'Bonding' is the arrangement of bricks in a definite pattern in order to provide a pleasing appearance but still maintain adequate strength for the job being constructed. The dimensions of bricks have been carefully designed to facilitate the bonding process.

Bricks must be lapped over each other in successive courses, both along the wall and across its thickness, to allow loads, including the wall's own weight, to be evenly distributed throughout the thickness and height of the wall, down to the foundations below. If bricks are not lapped over, straight vertical joints will occur. These will weaken the wall and prevent the even distribution of loads.

The most common bonding arrangement for brickwork is half-bond or 'stretcher bond'. The face bond consists of all stretchers except at return angles and stopped-ends, where a half-brick or 'bat' is introduced on alternate courses in order to maintain the half-brick overlap.

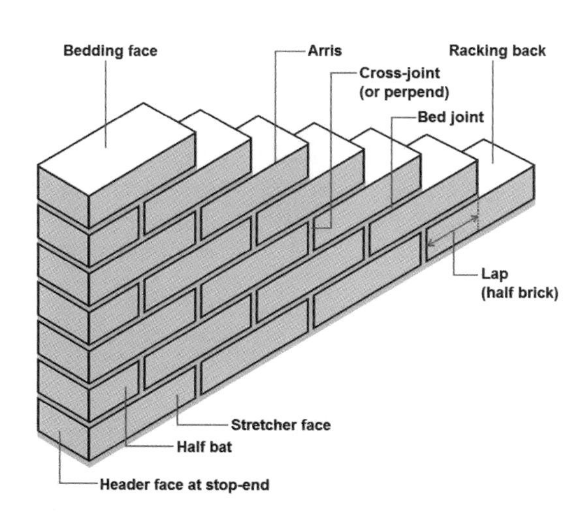

Fig. 83 Bonding terms and terminology relating to a stretcher bond wall.

Wall thicknesses are usually stated in brick sizes – half-brick thick, one-brick thick, and so on. Stretcher bond is most commonly used for half-brick walls such as the external leaf of cavity walls.

LOAD DISTRIBUTION

One of the main functions of a brick wall is to carry loads down to the foundations. This may be the load of the wall itself, which is known as 'dead load', or it may be loads that are carried by the wall, which are called 'superimposed loads'.

If the bricks were bedded one on top of the other the load would pass down a single stack of bricks, resulting in settlement and cracking of the wall (*see* Fig 84). Any lateral (sideways) pressure on the wall would simply push over a single stack of bricks. If the wall is properly bonded – each course

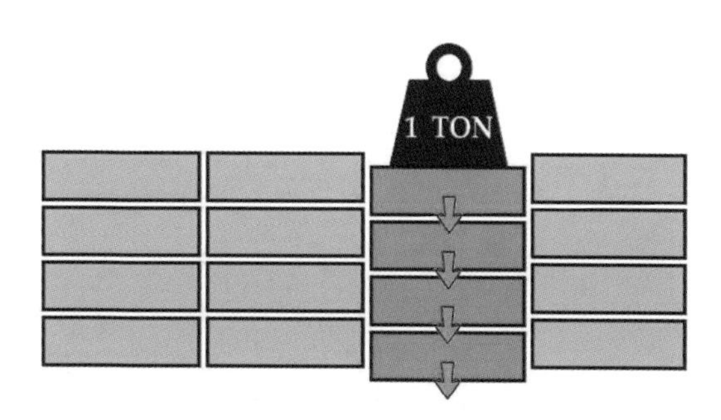

Fig. 84 Point load distribution on unbonded brickwork.

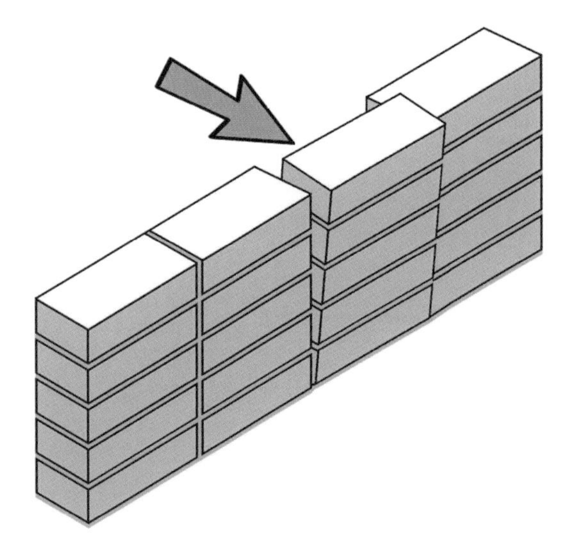

Fig. 85 Lateral load distribution on unbonded brickwork.

bedded with a half- or quarter-brick lapped over the course below – any lateral forces are distributed through adjacent bricks and the load is spread over a large face area of wall. The principle is the same in relation to a downward superimposed load (see Fig 86), where the bonding of the bricks is fundamental to the load distribution down the height of the wall.

STRETCHER BOND

Half-Brick Walls in Stretcher Bond

The term 'stretcher bond' is derived from the fact that all the bricks are laid with their stretcher faces showing. Half-brick walls in stretcher bond tend to be mostly used for the external leaf of a cavity wall. In addition and when strengthened by attached piers, they are used for garage walls, boundary and garden walls, and so on.

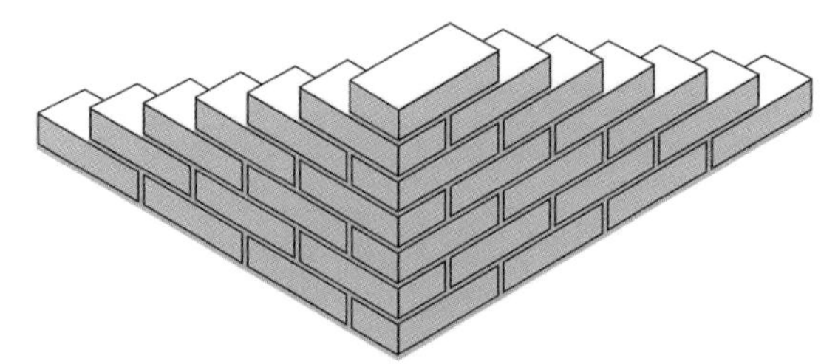

Fig. 87 Half-brick corner/return in stretcher bond.

Attached Piers in Stretcher Bond

'Pier' is another term for pillar and the purpose of an attached pier is to increase the strength of a wall against lateral (sideways) forces or throughout the wall's height where a point load (for example, from a steel beam) is exerted vertically on the wall.

Probably the most common application of attached piers is when constructing outbuildings such as garages. The inclusion of attached piers at key strength points, such as where doors are hung and also half-way down a long wall, provides sufficient strength to allow the use of half-brick walling in stretcher bond.

In order to provide adequate strength to half-brick walls, piers should be positioned no more than 3m apart along the length of the wall and no more than 3m away from return angles/corners, since the corner acts as a strengthening buttress.

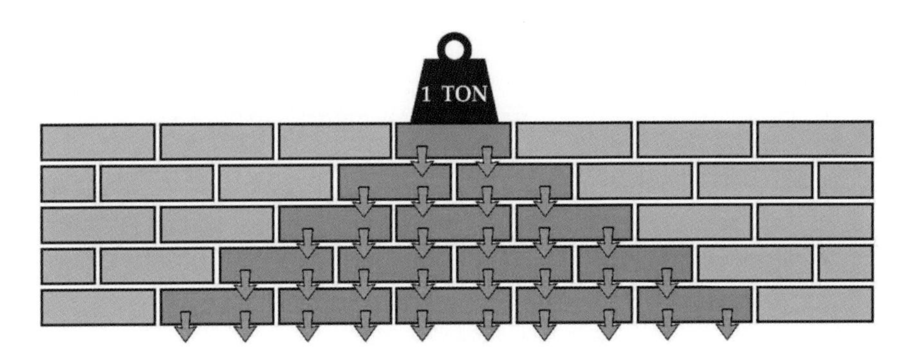

Fig. 86 Point load distribution on bonded brickwork.

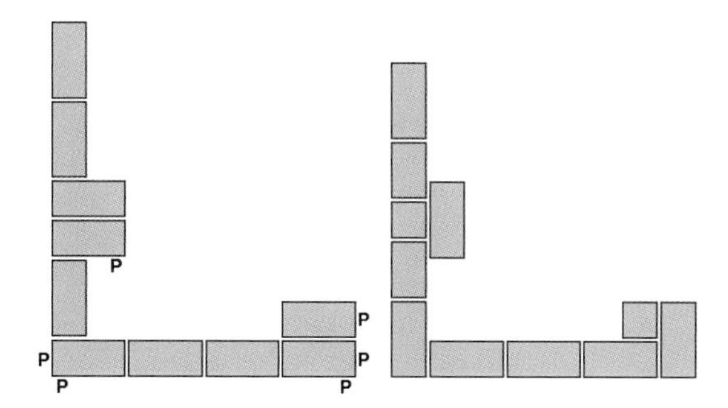

Fig. 88 Attached piers in half-brick stretcher bond.

It should be noted that it is usual for attached piers to be located on the back of half-brick walls in stretcher bond. Whilst the face sides of the wall and end piers can be plumbed vertically with a spirit level, the back and inside faces of the pier cannot. Dimensions of bricks will vary from one brick to the next, even from the same pack, and the biggest variance is in the length. Attached piers in stretcher bond require full bricks to pass through the full thickness of the piers on alternate courses. The variation in brick lengths means that only the sides of the piers indicated with a 'P' can be plumbed accurately, along with the face side of the external corner (Fig 88). The remaining sides of the piers should be 'lined-in by eye' to get as smooth a vertical line as possible. With this in mind, it is important to be selective about the bricks used in constructing such piers; any brick that is very different from the others should be discarded and another selected.

ONE-BRICK WALLING

One-brick walling – that is to say, walls that are one brick (215mm) thick – are no longer used for the external walls of houses and domestic buildings. Instead, their use tends to be limited to garden walls, retaining walls and boundary walls. Some bricklayers still refer to wall thicknesses in imperial terms, so one-brick walls are sometimes called '9-inch walls' and half-brick walls '4½-inch walls', based on the old brick sizes.

Stretcher bond has an inherent weakness in one-brick walling, due to the continuous vertical 'collar joint' that runs the full height of the wall between the front half-brick thickness and the back half-brick thickness. In some instances, one-brick walls are still built in stretcher bond, with the strength issue (or rather the lack of it) being addressed by treating the wall like a cavity wall. A 10mm cavity is left in between the two skins of stretcher bond, which are then tied together with short cavity-wall ties. For advice on the principles of cavity walling, see Chapter 11.

The stronger bonds used for one-brick walls make use of headers that are laid through the thickness of the wall in order to give the wall greater strength and improved load distribution through both height and thickness. The most common bonds for one-brick walls are English bond and Flemish bond which, due to the introduction of headers, are quarter-lap bonding arrangements.

Plumbing One-Brick Walls

The presence of headers laid through the thickness of the wall and the variation in the length of bricks means it is only possible accurately to plumb the front, stopped-end and return angles (corners) of one-brick walls built in English or Flemish bond. It is not possible to plumb the back accurately and this explains why the rear of such walls can have an uneven face plane. This is why one-brick walls are sometimes built in stretcher bond, in order to get as close as possible to a fair face on both sides, although, from a strength point of view, the use of this type of bond is not recommended.

English Bond

English bond comprises alternating courses of headers and stretchers. This bond is very strong, as alternating complete courses of headers eliminate any continuous internal vertical straight collar joints within the wall. Accordingly, an internal collar joint only appears on alternate courses between the front and back of the stretcher courses.

English bond tends to have quite a monotonous, repetitive appearance and is most commonly used where strength is favoured over aesthetics, such as for retaining walls.

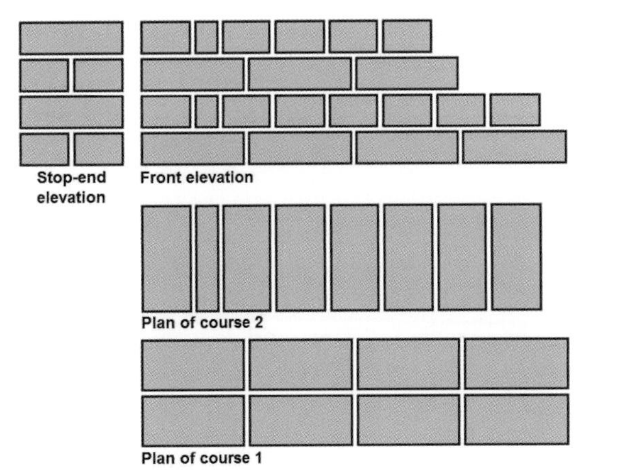

Fig. 89 English bond.

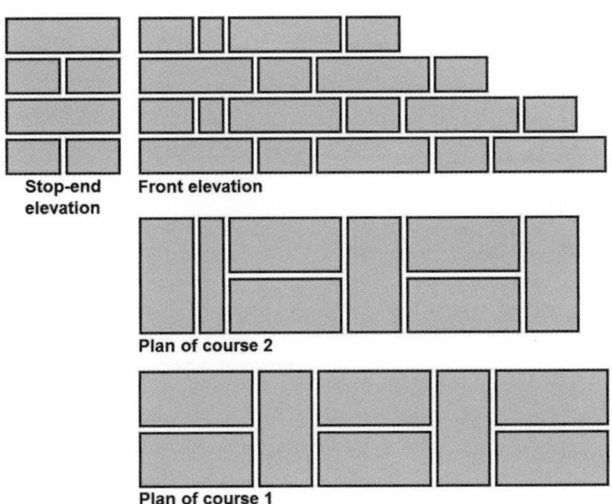

Fig. 90 Flemish bond.

Flemish Bond

Flemish bond, sometimes known as 'Dutch bond', comprises alternating headers and stretchers in the same course. While it is not by any means a weak bond, it is not as strong as English bond because internal vertical straight collar joints in the wall's height are not completely eliminated. Flemish bond is more pleasing on the eye than English bond so it tends to be used where a decorative appearance is more important than just strength.

Queen Closer

Both English and Flemish bonds are referred to as 'quarter-bonds' because the overlap of bricks from one cross-joint to the next is a quarter-brick. The quarter-bond is achieved by introducing a 'Queen Closer' next to the quoin or stopped-end header on alternate courses.

Fig. 91 Queen Closer.

A Queen Closer is cut from a header and is 46mm wide on face. In the same way that two headers, with an allowance for a 10mm cross-joint in between, are equal to the length of a stretcher, two quarter-bricks with an allowance for a 10mm cross-joint will be equal to a header – in other words, the width of a brick less a 10mm allowance for a cross-joint, then divided by two. For example, 102.5 – 10 = 92.5 and 92.5 ÷ 2 = 46.25mm; this would be rounded down to 46mm as it would be impossible to cut bricks on site to an accuracy of 0.25mm.

Practically, as a good guide to get the right size gap (2 × 10mm cross-joints + 46mm = 66mm) for a Queen Closer, place a dry brick on edge tight up to the end or quoin header. Lay the next brick to the other side of the temporary brick-on-edge, and tight up to it. Remove the brick-on-edge and a gap of 65mm will remain. While this is technically 1mm too narrow, the loss of 0.5mm in each cross-joint either side of a Queen Closer is of little or no consequence.

English and Flemish Bond 'Quoins' (Corners) and Junctions

On external return angles (corners) in English and Flemish bonds (*see* Figs 92 and 93), it is the tie

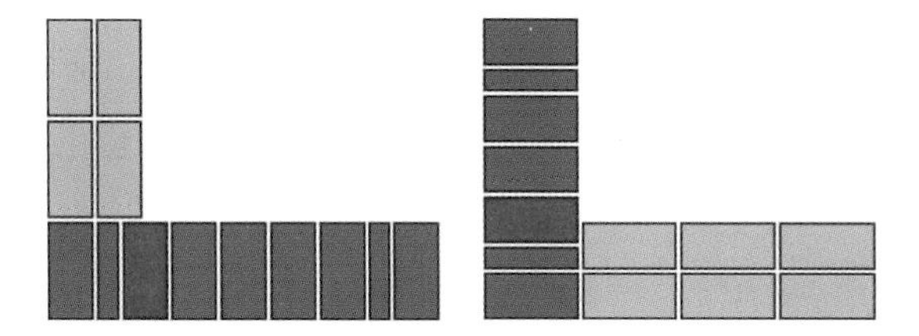

Fig. 92 English bond corner/return.

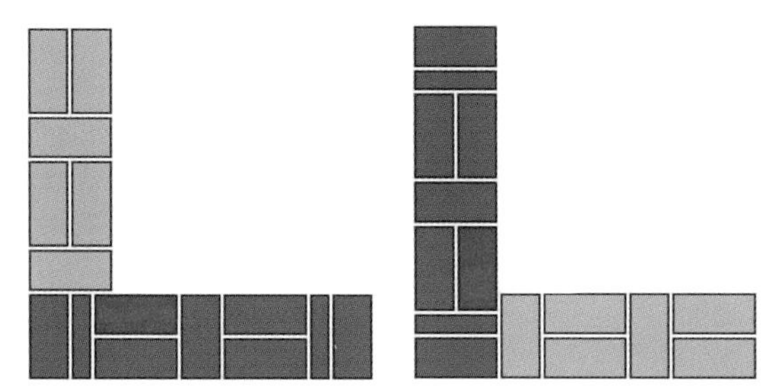

Fig. 93 Flemish bond corner/return.

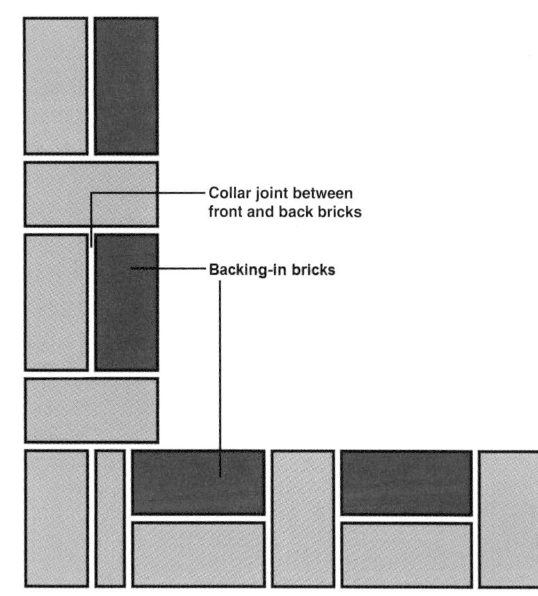

Fig. 94 Backing in bricks.

brick (identified in red) on every 'through-course' (identified in purple) that joins the wall together at that return angle. The other part of the return, which essentially stops at the back of the through-course, is known as the 'stopped course'. For clarity, both ends of the return walls are shown as stopped-ends utilizing a Queen Closer next to the end header on alternate courses. In addition, the through-courses are shown in a darker shade than the stopped courses.

When constructing one-brick walls, it is usual to lay all the bricks on the outer/face side of the wall on each course first and then 'back in' the stretchers that appear on the back of the wall. Fig 94 shows the bricks (in red) that would be laid last on each course. When backing in, it is not uncommon to lay the brick and to get it 'level' by merely ensuring it sits flush with the bricks around it – this is checked either by hand or by using the flat back of the trowel blade. When backing in, the bricklayer must be careful not to disturb the adjacent or surrounding bricks. For this reason, a cross-joint should not be applied to the back/common face of the brick (in other words, at the collar joint between the front and back bricks), as this could result in the front brick being pushed outwards. Instead, the collar joint is left empty and, when the whole course has been backed in, a trowel-full of mortar is 'fired' into the collar joint to fill it. This should not be done too vigorously as it may disturb the brickwork; there should be just enough force to fill the collar joint and nothing more.

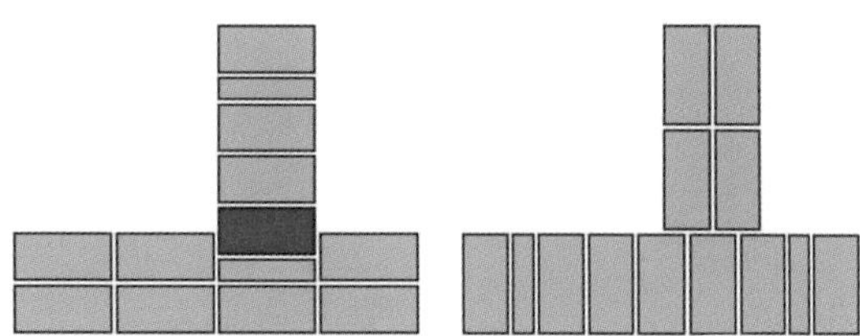

Fig. 95 English bond junction wall.

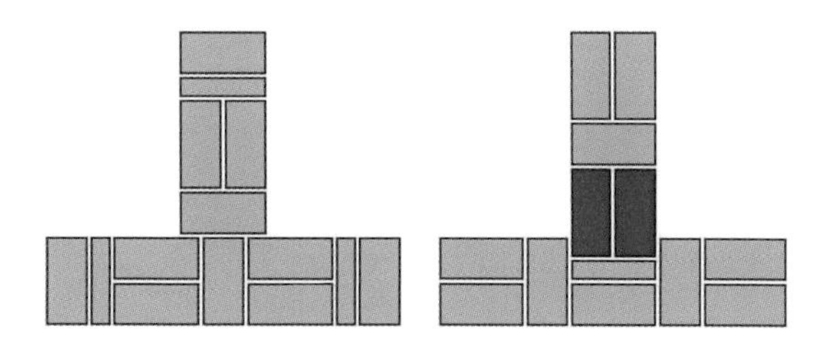

Fig. 96 Flemish bond junction wall.

Junction walls in English and Flemish bond follow much the same bonding principles as corners. Figs 95 and 96 show alternate courses for 'T' junctions in both English bond and Flemish bond with main tie bricks shown in red in each case. The tie bricks essentially tie in the junction wall to the main wall on alternate courses. In both cases, the Queen Closer placed at the heart of the junction on alternate courses enables the junction wall to be tied in and quarter-bond to be achieved on the junction wall. Again, for clarity, all three ends of both walls have been shown as stopped-ends utilizing a Queen Closer next to the end header on alternate courses.

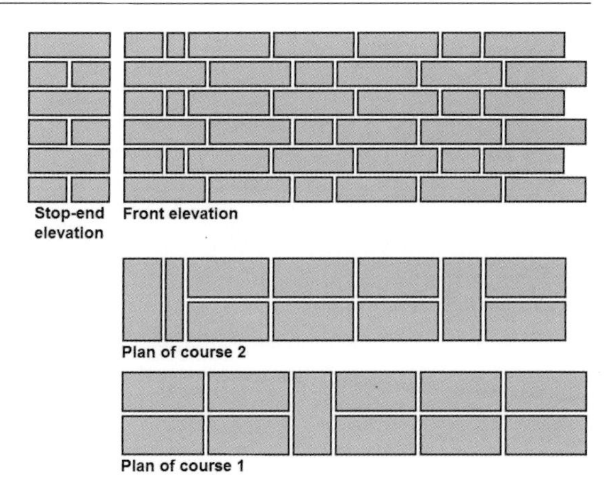

Fig. 98 Flemish garden wall bond.

GARDEN WALL BONDS

Bricks tend to vary in size and the most significant variation is usually in their length rather than any other dimension. In the case of one-brick walls in English or Flemish bond, the variation in length of the bricks used as headers means that, while an even and vertical face plane can be achieved on one side, it is impossible to achieve the same on the back. For this same reason, it is impossible to plumb up both sides of a one-brick wall.

To get over this, at least in part, particularly where both sides of a one-brick wall will be seen (such as on a free-standing garden wall), English and Flemish bonds are varied by reducing the number of headers. This forms a 'Garden Wall' variation of the bond.

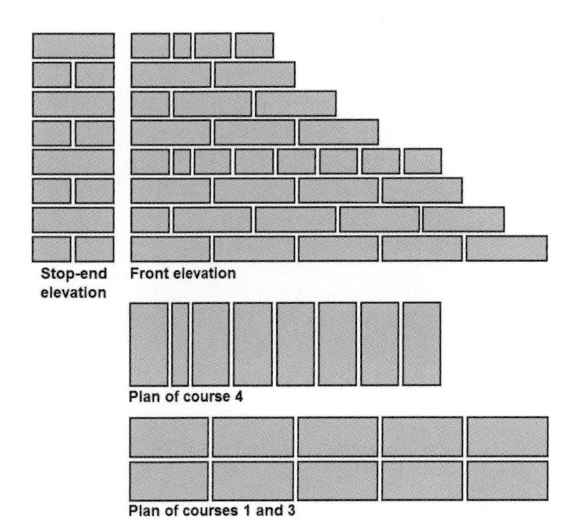

Fig. 97 English garden wall bond.

For English garden wall bond, the number of stretcher courses is increased to three or five between header courses, to improve the appearance of the rear face plane when both sides are seen. Strength is compromised due to the vertical collar joint that is formed between the front and back of the stretcher courses and runs the whole length of the wall.

For Flemish garden wall bond, the number of stretchers is increased to three or five between headers in each course, again to improve the appearance of the rear face plane. Compared to English garden wall bond, Flemish garden wall bond is stronger as the headers are more evenly distributed. This bond provides for a better compromise between strength and appearance.

MAKING ADJUSTMENTS

Dry Bonding

While it is desirable for the length of a wall to be designed to brick lengths, sometimes it will have to be of a length that does not 'work bricks', meaning that it will not be possible to lay full stretchers and/or headers. In such cases, cut bricks will have to be introduced to make the bond work to the length of wall required.

With this in mind, it is very important first to set-out a wall by laying bricks out with no mortar. This is called 'dry bonding' and will identify what

Adjusting Cross-Joints

When adjusting the width of cross-joints it must be done in such a way that all joints are adjusted to the same thickness in order that cross-joints remain vertically aligned and that the wall appearance looks uniform and consistent.

adjustments need to made to the bonding arrangement. Sometimes it will be necessary only to adjust the width of the cross-joints, however the tolerance is only +/– 3mm so joints can be a minimum of 7mm and a maximum of 13mm.

Reverse Bonding

If a wall works to half a brick length, stretcher bond can be reversed (*see* Fig 99), with half-bricks that form the stopped-ends appearing at alternate ends on alternate courses.

Broken Bonding

On stretcher bond walls that do not work to a half a brick length and for most other bonds, reverse bond is not an option so an alternative solution must be found. One possibility is to either widen or narrow the thickness of all the cross-joints throughout the length of the wall, to 'gain' or 'lose' the desired amount that means the wall length will then 'work bricks', whether to full- or half-brick length. The accepted tolerance for adjusting the width of cross-joints is 10mm + or –3mm; strictly speaking, this tolerance actually exists to allow for the slight variation in the sizes of individual bricks.

This course of action presents a number of problems. First, adjusting the width of cross-joints is generally not an option on short walls as the small number of cross-joints limits the scope for significantly adjusting the wall length. Second, wider cross-joints can look particularly unsightly, even when staying within the acceptable tolerance of +3mm, and applying a joint finish tends to make joints look wider still. On this basis, if cross-joints need to be adjusted then narrowing them is arguably the lesser of two evils. Third, the width of cross-joints must be consistent and the verticality of cross-joints throughout the height of the wall must not 'wander' unacceptably. The process of adjusting the width of cross-joints away from the usual 10mm tends to lead to a lack of consistency in width and 'wandering' cross-joints will inevitably appear, particularly on long walls.

Adjusting the width of cross-joints is best avoided, except where only very minor adjustments are required. The best option is always to

Fig. 99 Reverse bond.

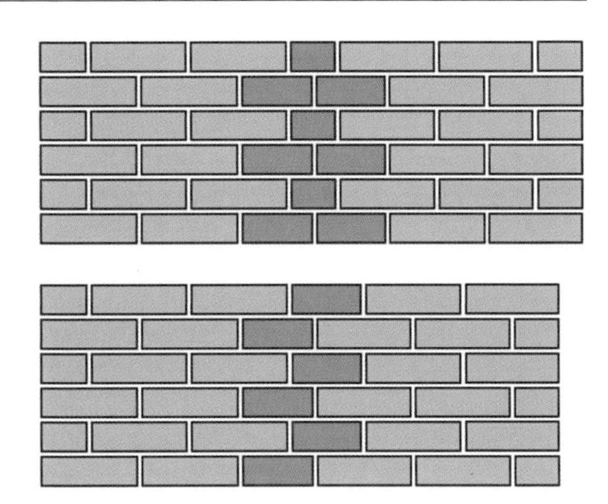

Fig. 100 Examples of broken bond.

Broken Bond or Reverse Bond?

A broken bond in the middle of the wall is also an alternative to a reverse bond on stretcher bond walls that work to half a brick length. Some bricklayers still prefer to use a reverse bond as it maintains the half-brick overlap where a broken bond does not.

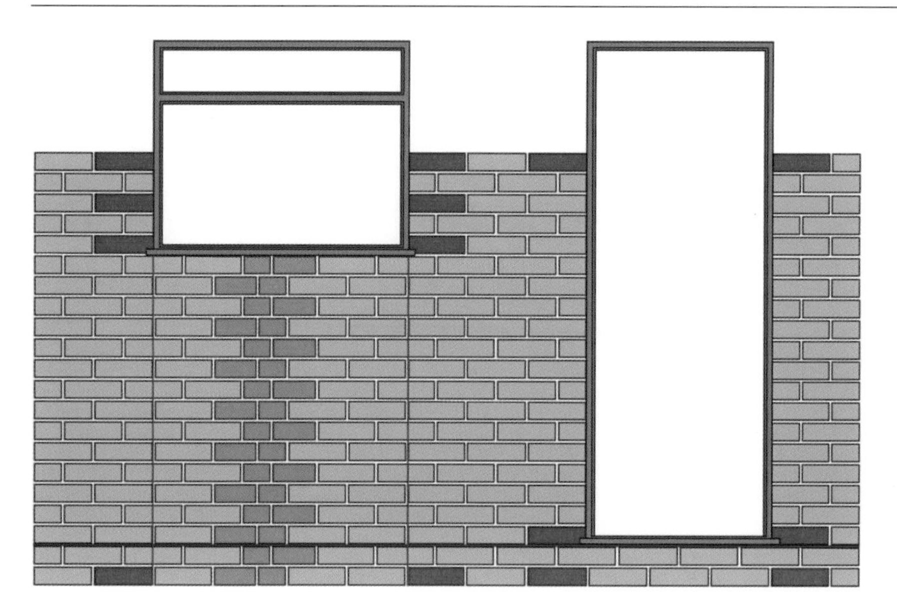

Fig. 101 Broken bond at openings.

dry bond the wall first, and then to take a view on whether the proposed adjustment is aesthetically acceptable. Where it is not acceptable, it becomes necessary to introduce cuts into the middle of straight walling. This is known as a 'broken bond'.

Broken bonds must visually and structurally maintain the integrity of the original bond. The smaller the cuts, the more noticeable they become, so the accepted bonding rule is that cuts no smaller than a half-brick should be introduced into the middle of a wall. This will maintain a minimum overlap of a quarter-brick. If it is any less, the structural capabilities of the wall in terms of load distribution will be impaired.

Maintaining the verticality of perpends or cross-joints has already been identified as an important factor in bricklaying generally, but extra care must be taken where the positioning of broken bonds is concerned. Excessive vertical deviation of cross-joints will draw undesirable attention to the broken bond.

When cutting bricks for a broken bond, they must be cut accurately and neatly (a fair-faced cut), and all cuts must be the same size. To this end, it is common for all the bricks required for a broken bond to be measured, marked and cut first in one batch. This is also a more efficient way of working, as the bricklayer does not have to stop on every course to cut bricks.

There will always be a few alternative ways of broken bonding to suit a specific situation and the bricklayer must find the most appropriate solution in each case. This is achieved by attempting to balance a number of factors: using cuts that are as big as possible so that the broken bond is not too noticeable; using as few cuts as possible on each course, so as to minimize the amount of time spent cutting; and, finally, ensuring that the original bond is not compromised aesthetically or structurally. This can be achieved only by dry bonding the centre portion of the walling, in order to test the various solutions available.

On buildings, any broken bonds should be set-out so that they are centred above and below any openings. This avoids any unsightly small cuts being placed either side of an opening at the reveals, helps to minimize the extent to which the broken bond is visible, and reduces the amount of cutting. The position of window and door reveals should be set-out with full stretchers on the first course of brickwork at finished ground level, which are then plumbed up the face of the wall as it is built in order to accurately position the frames at their required height. Fig 101 shows, in red, the position of the openings set-out in full stretchers at ground level and how these positions are transferred vertically to the reveals. There is a broken bond placed centrally under the window opening.

BONDING BLOCKWORK

Principles

Despite the difference in size compared to bricks, the principles of bonding blockwork are much the same as those for brickwork. A regular bond pattern should be maintained throughout a wall's length, ensuring a minimum overlap of a quarter of a block, although blockwork is almost always laid half-bond.

At stopped-ends, a half-block (215mm) is cut and introduced on alternate courses to maintain the bond. To establish half-bond at junctions and returns, a 105mm block bat should be built in adjacent to the return block (see Fig 102).

An alternative method is to introduce a brick-on-end with its bedding face turned outwards, but this is very bad practice. It is unsightly on fair-faced work and, being a completely different material, will move disproportionally to the blockwork, resulting in cracking. The only viable alternative method is to make use of purpose-made quoin-return blocks, which are 'L'-shaped on plan. This will ensure a half-bond at the corner and eliminate the need for any cutting. That said, the time saved by avoiding cutting blocks at corners will probably not justify the additional cost of such special blocks on smaller projects.

Bonding Junction Walls in Blockwork

Junction walls at 90-degree angles ('T' junctions) are quite common in blockwork, particularly when forming new internal partitions. A typical bond-

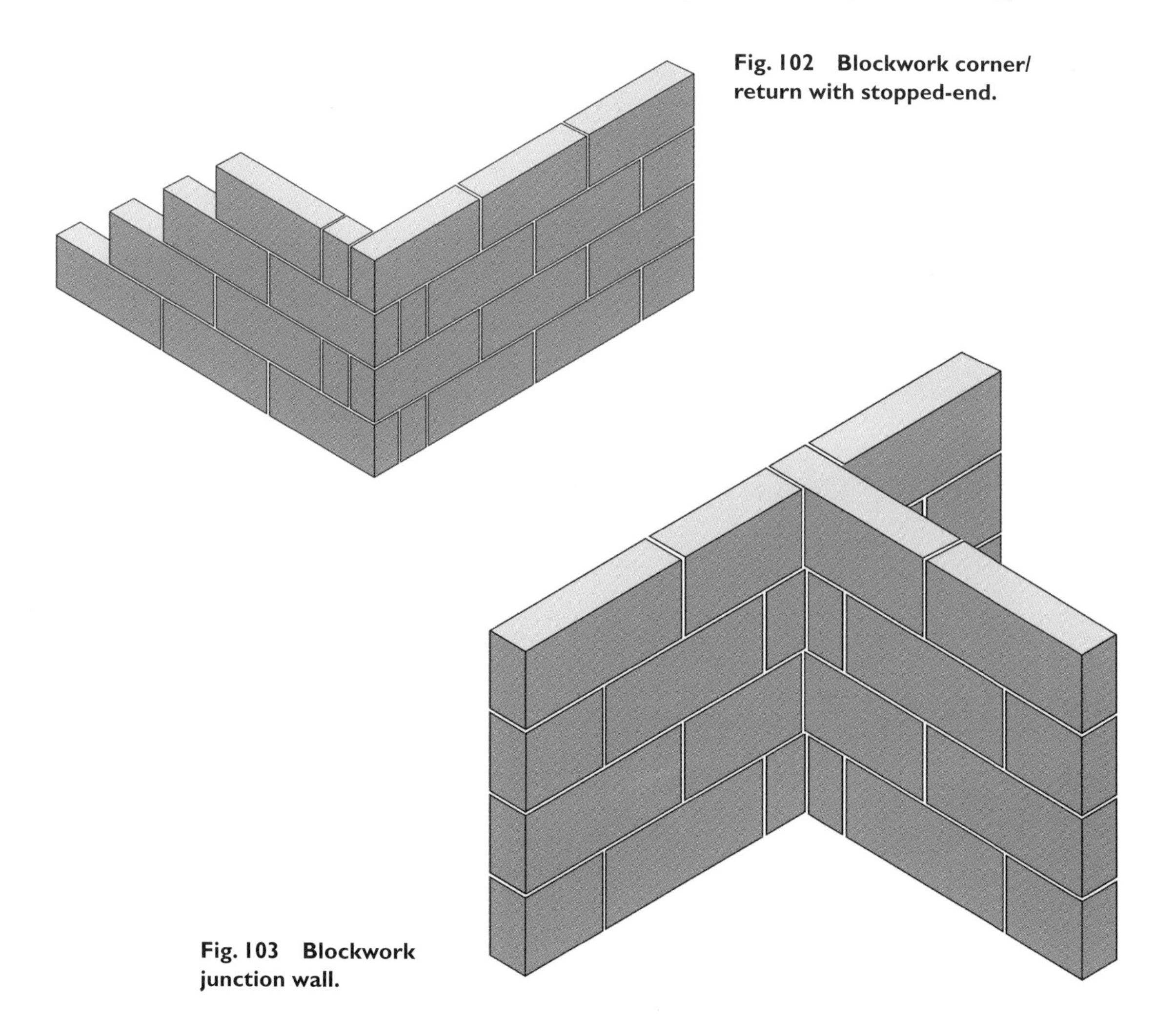

Fig. 102 Blockwork corner/ return with stopped-end.

Fig. 103 Blockwork junction wall.

ing arrangement for tying in the junction wall (see Fig 103) involves a three-quarter block (330mm) introduced into the main wall on alternate courses, which effectively leaves a 120mm indent (100mm block + 2 × 10mm cross-joints = 120mm). This allows the junction wall to be bonded into the main wall. On alternate courses of the junction wall a 105mm cut is introduced at the junction to maintain half-bond.

Attached Piers in 100mm Blockwork

Attached piers in blockwork are mostly required internally when there is a requirement to support the end of a lintel or structural beam. The attached pier in Fig 104 is 440mm wide and projects from the face of the main block wall by 225mm. The only way to tie in the pier to the main blockwork on alternate courses is to lay pairs of blocks flat on every course. On the courses where the blockwork is not tied in to the main wall, a full block is run through the back of the pier – this is simply a reverse of the bonding arrangement on the tie-courses where a full block is visible at the front of the pier.

Fig. 104 Attached pier in blockwork.

KEY POINTS TO ENSURE GOOD BONDING

- Carefully set-out walls and any broken bonds, and dry bond whenever possible.
- Bricks must be lapped over each other in successive courses, both along the wall and across its thickness.
- Overlapping allows loads, including the wall's own weight, to be evenly distributed throughout the thickness and height of the wall down to the foundations below.
- If the bricks are not lapped over each other 'straight joints' will occur, which are unsightly and considerably weaken the wall. External 'straight joints' must be avoided at all times.
- Internal 'straight joints' through the height of the wall must be eliminated or kept to a minimum. This is why one-brick walls in stretcher bond should be avoided.
- Closers should only be placed next to quoin/stopped-end headers and nothing smaller than a half-brick should be placed in the middle of a wall.
- Bats and closers must be cut accurately to maintain the bond and verticality of cross-joints.
- Cross-joints in alternate courses should be kept vertical. Cross-joints should not be allowed to 'wander', which comprises the overlap of the bricks – when this happens excessively it is known as 'losing the bond'.
- All mortar joints must be full as any voids will weaken the wall.

CHAPTER 9

Bricklaying Basic Skills

LOADING OUT THE WORK AREA

Before bricklaying commences, it is important to consider where to position the materials and where to stand. The aim is to eliminate as much excessive bending and movement on the part of the bricklayer as this is tiresome and slows down the work. The positioning of materials ready for bricklaying is called 'loading out' or 'stacking out'. They must be placed to ensure efficiency and economy of movement for the bricklayer, with everything in easy reach.

First, a rough calculation of the bricks and/or blocks required for the job should be carried out and the total number of bricks and/or blocks spread out evenly in neat, bonded stacks along the length of the wall to be built. Bonding the bricks or blocks at this stage will reduce the risk of them being knocked over and creating a safety hazard. Also, wherever possible, the faces of the bricks should be turned away from the mortar board so they do not get splashed and stained.

On long walls it is good practice for there to be a number of mortar boards (or 'spots') positioned conveniently, so that the bricklayer is always within easy reach of both bricks and mortar. Mortar boards are lifted up on four bricks, one at each corner, which helps to keep the work area tidy but also reduces the amount of bending for the bricklayer and the travel distance for the mortar. This may seem like a small time-saving but it adds up considerably over the course of a working day.

Fig. 105 Loading out the work area.

Mixing Bricks When Loading Out

Bricks are usually delivered in packs which are banded in plastic and shrink-wrapped in polythene. Depending on supplier, the number of bricks in a pack can be between 390 and 525, but 456 bricks or 475 are typical pack sizes. Being made from a natural material, there is always the distinct possibility that facing bricks will vary in colour or shade from one pack to the next. This could be due to variations in temperature within the kiln during the firing process, or to bricks in different packs being from different firings. On large jobs that make use of multiple packs it is important, when loading out, to mix bricks from at least three different packs (if possible), otherwise the finished brickwork could suffer from horizontal 'banding' of different shades.

Stacks of bricks and the front of the mortar board should be placed approximately 600mm from the face of the wall being built, to give adequate and safe working space but avoiding the bricklayer having to engage in excessive movement between the wall and materials. When loading out, bricks should be carefully selected in order to avoid chipped and cracked bricks appearing on the face of the wall.

Typically, mortar boards are simply made from an off-cut of 20mm plywood (or similar), cut to a usable practical size of around 750mm × 750mm. Mortar boards should be dampened before being loaded out with mortar, as a dry board will absorb water from the mortar mix. This will dry out the mortar and reduce its workability.

Finally, and before any bricklaying starts, any adjacent paved areas should be protected with polythene sheeting to prevent them being contaminated or stained with mortar splashes. Tarmac, for example, is notoriously difficult to clean when mortar has been dropped on it and trodden in! Even with sheeting down, paved areas should still be hosed down at the end of every working day.

BASIC BRICKLAYING TECHNIQUE

The basic technique of bricklaying, from picking up a trowel full of mortar to the point of bedding bricks, involves a number of distinct steps and processes. Each bricklayer will bring a slightly different personal style to the way that these hand skills are carried out, but it is fair to say that, regardless of the style used, the most important issue is the effectiveness of the end result, provided that the following basic techniques are employed.

Cutting and Rolling Mortar

The mortar is placed towards the back of the mortar board, with a clear space left at the front of the board for the process of 'cutting and rolling', which is the precursor to picking up a trowel full of mortar. The correct working position for picking up the mortar is side on to the mortar board with the 'trowel hand' over the board and the 'laying hand' over the wall being constructed. There is a very good reason for adopting this particular working position, which will become clear later.

Cutting and rolling is carried out as follows:

1. Hold the brick trowel with your thumb in line with the handle and pointing towards the tip of the blade. In general terms, the trowel never leaves the trowel hand!
2. Cut away from the main pile a quantity of mortar that is roughly sufficient to fill the trowel blade.

Fig. 106 Correct method of holding the trowel.

Fig. 107 Cutting mortar.

Fig. 108 Rolling mortar.

Fig. 109 Preparation for picking up the mortar.

Fig. 110 Picking up the mortar.

3. Using a 'sawing' motion, roll the mortar across towards the front of the board to form a roll big enough to cover the trowel blade. The sawing motion must not be too straight as this will produce a roll of mortar that is too long for the trowel to pick up. Instead, alter the angle of the trowel to draw the mortar in at each end.
4. During the cutting and rolling process, keep the trowel blade in contact with the board so it cleans the board as the mortar is rolled. The blade is not kept completely perpendicular to the board; instead, it is inclined over slightly towards the front of the board to assist the rolling process. The aim is to cut and roll a piece of mortar that is much the same shape and plan-size as the trowel blade.
5. Having cut and rolled the mortar, move the trowel back from the roll of mortar and turn it so that the blade is horizontal about 1mm above the spot board and around 50mm away from the mortar roll, in preparation for picking it up. At this point, some bricklayers prefer to loosen their grip and turn the trowel slightly outwards to avoid having to bend and strain the wrist too much. Again, this is a matter of personal style and not an issue of technical correctness.
6. With a sharp movement and leading with the long inside edge of the trowel, pick up the roll of mortar. This movement needs to be fast enough to overcome the inertia of the roll of mortar but not so fast that the trowel flies straight through it!

Spreading Bed Joints

Having achieved a trowel full of mortar, the next stage is the spreading of a bed joint, followed by furrowing ready to receive bricks. Bed joints must not be spread too thickly as too much effort will be required to tap the brick down to gauge (a 10mm bed joint). Ideally, a bed joint should be spread sufficiently thinly so the brick can be pressed to gauge with little or no tapping from the trowel.

At this point the bricklayer should not turn round so that the trowel hand is over the wall, but bring the trowel hand across the body, and then proceed as follows:

1. Hold the trowel of mortar over the position where the bed joint is required. This may be the surface of a concrete foundation at the start of new work or on top of the previous course of brickwork for work that is ongoing.
2. Spreading involves transferring a trowelful of mortar into something resembling a line of mortar. Turn the trowel vertically through 90 degrees and combine a dropping motion with, simultaneously, sweeping or drawing the trowel blade backwards along and parallel to the line of brickwork. In general terms, the direction of spreading is always along the wall and not across it, regardless of whether headers or stretchers are being laid.
3. If the mortar bed has not quite spread evenly, pick up thicker parts with the trowel edge and deposit them at one end of the line of mortar so that the initial spreading is more even and elongated.

Fig. 111 Spreading mortar.

Fig. 112 Elongating the initial spread of mortar.

Fig. 113 Furrowing mortar.

4. With the point of the trowel, 'furrow' the spread mortar along its length and backwards with a series of undulating trowel movements. This creates an indentation along the spread mortar, which, when a brick is pressed into it, fills to form a full bed joint. It avoids too much mortar being squeezed out of the joint, which could fall down the face of the brickwork below and cause staining. Full joints are vital, and furrowing the mortar too deeply can result in voids in the bed joints, which will affect the structural strength of the finished brickwork.
5. Having furrowed the mortar, use the inside edge of the trowel with the top face of the blade turned outwards to trim off any mortar that overhangs front and back in order to provide a neat edge to the bed joint in readiness to receive bricks.

Handling Bricks

The hand not holding the trowel is referred to as the 'laying hand'. The correct way to hold the brick in this hand is across the width. Holding a brick in this way allows it to be spun in the laying hand to the correct face or bedding plane before laying or before applying a 'cross-joint' to one end in readiness for the brick to be laid adjacent to another.

Fig. 114 The correct way to hold a brick.

When spreading bed joints for cavity walling it is very important not to let the mortar bed overhang on the cavity side. Great care should be taken to prevent mortar falling down the cavity, which could result in the cavity becoming bridged and lead to water ingress through the finished wall. The mortar that is trimmed off should not be allowed to fall or be deposited on the floor. Instead, efficient, practical use should be made of it by placing it at one end of the line of mortar so that the initial spreading is further elongated.

Fig. 115 Laying a brick.

Bedding Bricks

Having successfully spread a bed joint, the bricks are then bedded as follows:

1. Lay the brick by placing it on to the mortar bed and gently press it until a 10mm bed joint is achieved (in other words, 'to gauge').
2. 'Eye down' the cross-joints for verticality with those below, and correctly align the brick, by eye, as close as possible to its final position.
3. Using the inside edge of the trowel, with the top face of the blade turned outwards, cleanly trim off the mortar that squeezes out front and back. Do not turn the trowel inwards when trimming off as this will smear mortar up the face of the brick that has been laid. This is particularly important on the face side of the wall. Facing brickwork should be kept as clean as possible, with minimal smudging and no visual defects.

Fig. 116 Trimming off excess mortar from the bed joint.

'Eyeing Down'

Eyeing down the cross-joints with those below is one of the key reasons for the bricklayer's stance, which means that the trowel hand remains over the wall being built. The act of laying the brick causes the bricklayer to lean towards the wall and this automatically draws the eye line down through the vertical line of cross-joints. This makes it easier to spot if the cross-joints are beginning to 'wander' out of vertical alignment.

Again, mortar that is trimmed off should not be allowed to fall or be deposited on the floor. Instead, if there is enough of it, it can be used for the cross-joint on the next brick. This will save time, reduce movement on the part of the bricklayer and improve efficiency.

Application of Cross-Joints

The end of a brick that has the cross-joint applied depends on whether a bricklayer is moving forwards or backwards along the wall as it is built. When moving backwards along the wall, the front end of the brick, as it is held, will have the cross-joint applied. The opposite applies when moving forwards. The latter is slightly more awkward as it is necessary to twist the wrist of the laying hand round in order to apply the mortar joint at the other end of the brick. For this reason, most bricklayers prefer to work backwards along a wall.

To apply a cross-joint, the procedure is as follows:

1. Start with a small scoop of mortar on the end of the trowel. Flick the wrist of the trowel hand slightly but firmly, to cause the mortar to spread across and adhere to the trowel blade.
2. Pick up the brick across its width with the thumb of the laying hand resting on the face side of the brick. Hold the brick and the trowel slightly apart in front of the body in preparation.
3. Apply the first part of the cross-joint by drawing the trowel blade down across the front edge of the header face, so that a portion of the mortar from the blade is stuck or 'buttered' on to the brick. Take care not to smear or stain the face of the brick.
4. Apply the second part of the cross-joint by reorienting the brick in relation to the trowel and drawing the blade down across the back edge of the header face. Again, a portion of the mortar from the blade should stick on to the brick.
5. Reorient the brick in relation to the trowel again and draw the blade down across the bottom edge of the header face to complete the cross-joint.

Fig. 117 Preparing to apply a cross-joint.

Fig. 118 Applying the first part of the cross-joint.

Fig. 119 Applying the second part of the cross-joint.

Fig. 120 Completing the cross-joint.

The brick is now 'buttered' on the bottom and both side arrises, at its header end, ready for bedding. Sufficient mortar should be used to ensure a full cross-joint when the brick is pressed up to an adjacent brick, since voids will affect the strength of the finished wall. It is not necessary to butter a fourth portion of mortar to the top edge of the header face, as the bed joint of the next course that gets spread over the top will fill it up. Voids in cross-joints create weaknesses in the finished wall, so it is vital to ensure that the three edges are buttered with mortar. In actual fact, the order in which the edges are buttered does not matter, nor does it matter whether the brick is moved in relation to the trowel or vice versa. These are issues of personal style with the only proviso being that the cross-joint must be full and the face of the brick kept clean.

Cross-joints that are too thick will make bricks difficult to lay against adjacent bricks, with the possibility of the cross-joints ending up too wide. The excess effort required to achieve a 10mm cross-joint could also affect the adjacent brickwork, particularly if the adjacent brick is the 'quoin

Fig. 121 Laying the next brick.

brick' at the end or corner of a wall, which is easily disturbed. The amount of mortar should be sufficient, ideally, for the brick to be pressed into place without the need for the other end of the brick to be tapped with the trowel.

Cross-Joint (Perpend) Width and Vertical Deviation

The terms 'perpend', or 'perp', is derived from the fact that vertical joints are perpendicular to the bed joints. Cross-joints or perpends should be of a consistent width throughout the wall and a standard cross-joint is 10mm wide. In practice, however, they sometimes have to vary in size to allow for the slight variation in the sizes of individual bricks (or sometimes at the bricklayer's discretion to accommodate wall lengths that do not quite 'work bricks'). An 'industrial tolerance' of + or –3 mm is considered adequate in dealing with the former, meaning that cross-joints can be as narrow as 7mm or as wide as 13mm and still be deemed acceptable. However, wide cross-joints can look particularly unsightly even when staying within the acceptable tolerance of +3mm, and applying a joint finish tends to make joints look wider still. On this basis, if cross-joints need to be adjusted, it is probably more acceptable to narrow them.

The vertical alignment of cross-joints must not 'wander' – a deviation from the vertical of only 5mm in the height of the wall is deemed acceptable. The process of adjusting the width of cross-joints away from the usual 10mm tends to lead to a lack of consistent width and 'wandering' cross-joints will inevitably appear, particularly on long walls. Vertically eyeing in the cross-joints as the bricks are laid will undoubtedly help, but it is also good practice to plumb cross-joints at regular intervals along the face of the wall. If the cross-joints are allowed to wander away from a true vertical, the appearance of the finished brickwork is detrimentally affected (*see* Fig 122). It must be acknowledged, however, that significant dimensional deviations in the bricks can make it very difficult to obtain a consistent 10mm cross-joint and perfect vertical alignment! Sometimes, the only way forward is to dry bond the proposed wall carefully, and make every effort to maintain the best compromise between aesthetics and staying within the accepted tolerances. It is a question of striking a balance between what is right and what looks right, given the limitations of the materials being used.

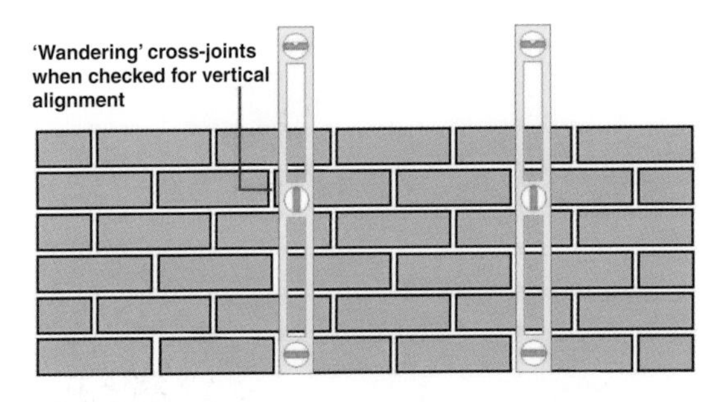

Fig. 122 Example of 'wandering' cross-joints.

> **Industrial Tolerances**
>
> The construction industry works to standards known as 'industrial tolerances'. These are in effect allowable 'margins for error', beyond which the quality of the finished work is considered unacceptable. Industrial tolerances for brickwork are designed to accommodate shortcomings in the materials themselves, since bricks are man-made from naturally occurring clay, and will therefore never be exactly the same from one brick to the next. That said, the leeway afforded by the tolerances does not mean that the bricklayer should strive for anything less than perfection where possible.

BASIC PRINCIPLES OF CONSTRUCTING WALLS

The four key features of successful brickwork are that it is constructed level, vertical (plumb), to gauge and in line. Bringing these four features together means adopting the fundamental principle that walling, even one short course of bricks, is constructed by first building the ends or corners of the wall plumb, level, and to gauge, and then filling in the brickwork in between, by 'running-in' (*see* Fig 123). The accuracy of the in-filled brickwork in terms of level, plumb and gauge can only be as accurate as the corners or ends of the wall. Any

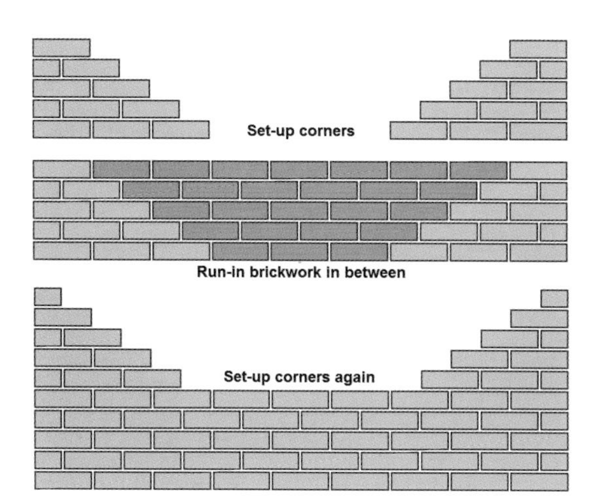

Fig. 123 First, the ends or corners of the walling are built and then the brickwork is run-in in between.

mistakes or inaccuracies in the corners or ends will be transferred to, and replicated in, the middle section of the wall when it is run-in in between.

This principle applies to all brickwork projects, from the largest to the smallest. After running-in, the process of building ends or corners is repeated on top ready for running-in again. Stepping the brickwork back as the end or corner is built is referred to as 'racking back' and is the most practical and common method of temporarily terminating brickwork. It allows it to be built on to later with no loss of strength where new brickwork meets old.

Brick for brick, when compared to running-in, building the ends or corners of walling is by far the most time-consuming aspect of bricklaying, because such care and attention must be given to levelling, plumbing up and ranging in every brick, to ensure accuracy. Accordingly, it is a false economy to set up large, high corners, as a disproportionate amount of time will be wasted on such activities. It is much more efficient in terms of time to build two or three small corners rather than one big one, as more time will be spent on the faster task of running-in. There are a number of methods for ensuring the accurate construction of corners and running-in, but first it is necessary to address issues of gauge, level and plumb.

CHECKING FOR GAUGE

Checking for Gauge Above Ground

The first brick on every new course (the 'quoin brick') must be checked and adjusted for gauge, where gauge is the vertical measure applied to the height of brickwork in terms of the combined thickness of bricks and the mortar bed joints between courses. All bed joints should be of regular thickness throughout the height of a wall – 10mm thick with an accepted industrial tolerance of + or –5 mm in any 3.0m height.

Bricklayers will usually check gauge by measuring with a tape, in multiples of the thickness of a course of bricks (65mm brick + 10mm bed joint = 75mm gauge). Typically, they will check that every fourth course measures 300mm. Specialist bricklayer's measuring tapes are available, marked out for the purpose of gauging brickwork.

A more traditional alternative is a timber gauge lath marked with 75mm graduations. The gauge lath is made from a length of timber 50mm × 20mm in section (or similar), with the 75mm graduations marked on with permanent, shallow saw cuts. Adjusting a brick to gauge is achieved by simply tapping the brick down with the outside edge of the blade until a 10mm bed joint is reached. When tapping the brick down, cultivating a good eye for keeping the brick as horizontal as possible will make subsequent levelling easier.

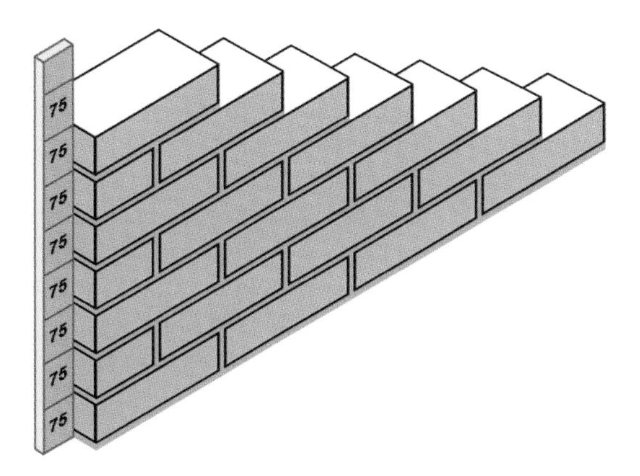

Fig. 124 Gauging brickwork with a gauge lath.

Adjustments to Gauge Below Ground

In theory, if both ends of a wall are erected to the same gauge from a horizontal, base level then every course of the wall will be level throughout its full height. In practice, however, a bricklayer cannot rely on a foundation being perfectly flat and/or level so a reference point must be established in the form of a datum level.

In terms of site practice for long walls, a datum level peg is fixed, typically, at the height of the horizontal DPC (on a building no less than 150mm above finished ground level) and this datum level is transferred to pegs at both ends or all corners of the wall or building being constructed.

For small walls or piers, where the length of the wall is within the limits of the length of a spirit level (or, at worst, a straight-edge), only one gauging point is required at one end. Level can be easily transferred from this, so only one datum peg needs to be positioned adjacent to the wall being constructed. Where a small wall or pier is completely isolated and remote from any other brickwork, the brickwork can simply be gauged straight off the foundation concrete as there is no practical or aesthetic requirement to marry up the gauge or level of the new work to anything nearby.

For details of the establishment of datum, its importance and methods of transferring levels, see Chapter 6.

The idea is that the brickwork, as it is built upwards from the foundation, finishes level at a point that coincides with the datum level. Any adjustments in terms of increasing ('picking up') or reducing ('grinding down') the thickness of bed joints to achieve this objective must take place below ground, where they will not be seen!

Once a datum peg has been transferred to each corner or end of the wall, a spirit level and gauge lath can be used to determine the extent to which gauge needs to be adjusted below ground (*see* Fig 125). Where the brickwork falls short of datum by less than half a course (37.5mm), it is easier to 'pick up' the thickness of the bed joints to make up the difference. This adjustment should be gradually spread over a number of courses rather than being done with a couple of huge bed joints, since joints that are overly thick could squeeze out or sink with the weight of any brickwork built imme-

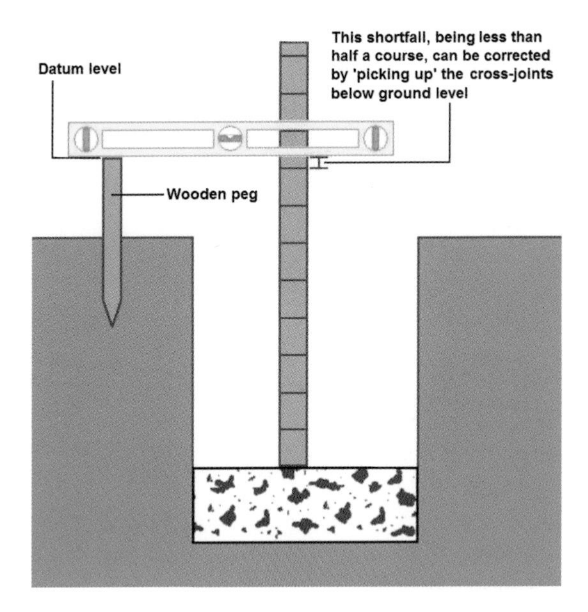

Fig. 125 Checking gauge between datum and foundation level.

diately on top. For this reason, where the shortfall exceeds half a course – for example, where it is around 50mm – it is better to add in another course of bricks and 'grind down' the joint thickness to accommodate the extra course, rather than 'picking up'.

A simple strip foundation would require a minimum thickness of 150mm and assume a distance between the top of the concrete to finished ground level of 1000mm. Assuming that datum has been established at DPC level of 150mm above finished ground level, the overall distance from datum to the top of the foundation concrete is 1150mm below datum. This figure (1150mm) divided by a brickwork gauge of 75mm works out to 15.33 courses of brickwork from the top of the foundation to datum level, which clearly does not work to gauge – fifteen courses would finish short and sixteen would finish too high. Accordingly, the bricklayer will have to make adjustments to gauge by either 'picking up' the bed joints in fifteen courses to gain 0.33 of a course (approximately 25mm) as the brickwork is built out of the ground, or by laying sixteen courses instead of fifteen and 'grinding down' the joints by 50mm to lose 0.67 of a course.

The adjustments to gauge/bed joint thickness must be made below ground level where they are not seen, so no adjustment is possible in the two courses between finished ground level and DPC/ datum level. In practice, the bricklayer's choice is either to 'pick up' 25mm in thirteen courses or to 'grind down' 50mm in fourteen courses. The former will require every bed joint below ground level to increase in average thickness by a fraction under 2mm. This should not prove too difficult to achieve consistently and it is very unlikely that the slightly thicker bed joints will start to squeeze out as the brickwork is built higher. The latter option, on the other hand, requires each bed joint to be ground down/reduced in average thickness by 3–4mm. This would make for very thin bed joints and would be more awkward to achieve successfully. 'Picking up' the thickness of the bed joints slightly is clearly the easier and more practical option in this case.

Whichever method is adopted, the gauge should be regularly checked and re-checked as the brickwork is built out of the ground, to ensure that the bed joints can revert to the correct thickness of 10mm by the time finished ground level is reached.

A third alternative, albeit more involved, is to excavate the foundation trench to a precise, calculated level where the brickwork will work to gauge from the top of the foundation up to datum level without the need for adjusting the thickness of bed joints. For more on this, *see* Chapter 6.

Once the brickwork has reached the datum level – in this case DPC level – then a second peg can be inserted into the ground upon which the gauge lath can be positioned in order to gauge every course above DPC (*see* Fig 126). The ability to position the second peg assumes, of course, that the foundation trench has been back-filled. If this is not the case, a masonry nail can be inserted into the bottom of the bed joint at DPC level, and the gauge lath can be rested on this. This has a couple of disadvantages: the hole will need to be made good later, and the nail will cause the gauge lath to sit too high by a distance equal to its thickness. While this is a very small amount, some bricklayers prefer to

Accuracy When Gauging Brickwork

When gauging the first brick in every new course, if the brick is sitting high it should simply be tapped down. If it is sitting low, because the bed joint is too thin, it is wrong to carry on, thinking that the error can be rectified on the next course with a thicker bed joint. Instead, the brick should be picked up, a thicker bed joint re-spread and the brick laid again!

Gauging Door and Window Frames

The methods used to check and adjust gauge from foundation level are the same when building door or window frames into walling where the frame does not quite work precisely to the normal gauge of brickwork. Sometimes the gauge will have to be picked up or ground down so that the brickwork finishes level at the top of the frame, ready to receive a lintel on top. Given that any adjustments will be made in brickwork that will be visible, a great deal of care must be taken to ensure that the bed joint thickness is altered as uniformly as possible.

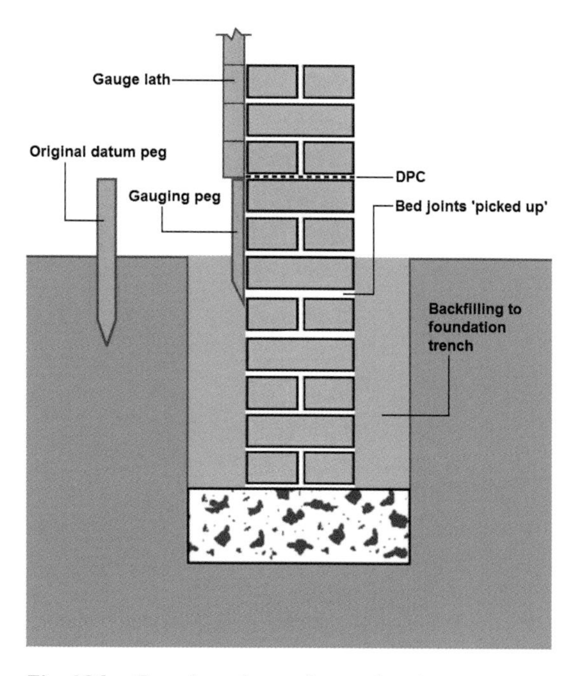

Fig. 126 Gauging above datum level.

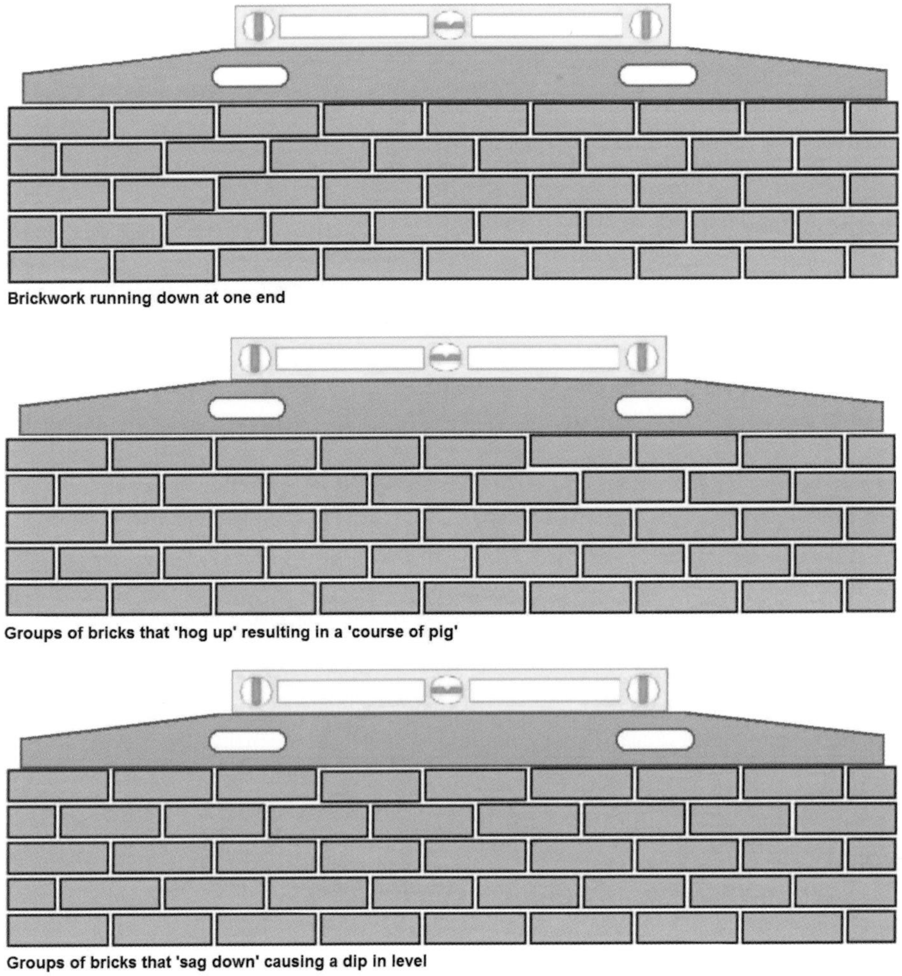

Brickwork running down at one end

Groups of bricks that 'hog up' resulting in a 'course of pig'

Groups of bricks that 'sag down' causing a dip in level

Fig. 127 Common defects when levelling brickwork.

make a small notch in the bottom of the gauge lath to accommodate the nail and eliminate any error.

LEVELLING AND PLUMBING BRICKWORK

Checking for Level

When laying bricks, a spirit level should be used to check that they are horizontal. Adjustments are made by tapping the top of the brick with the outside cutting edge of the trowel blade, not the end of the handle – using the handle will result in any mortar left on the trowel blade falling on to the trowel hand! Also try and keep the blade within the plan area of the brick being adjusted to level so that no mortar falls from the trowel down the face of the brickwork. When adjusting bricks to level, only tap along the centre lines of the brick so that the brick is only adjusted in one direction at a time.

All brickwork, in general, must be level with an accepted tolerance of + or – 3mm in a 2m length. The practical implication of this is that when a 2m long straight-edge is placed on top of the brickwork and held in a level/horizontal position there must be no gaps that exceed 3mm. Gaps may result from one end of the wall running up or down out of level, individual bricks or groups of bricks that 'hog up', causing a bump (sometimes known as a 'course of pig'), or individual bricks or groups of bricks that 'sag down', causing a dip.

Plumbing Brickwork

The ends or corners of brick walls must be built vertically upright, or 'plumb' on their face sides, and this is also checked by using a spirit level. Accurately 'plumbing up' the ends or corners is key to the basic principle of constructing walling. Having established fixed, vertical end points it is then possible to 'run-in' the brickwork to a spirit level or string-line (depending on the wall length), thus ensuring that the whole face plane of the wall is vertically aligned.

The acceptable industrial tolerance for plumb is + or – 3mm in 1.0 m height. Whilst brickwork may be generally vertical throughout its overall height, this is not enough on its own. Brickwork can still fail to meet the accepted tolerance for plumb because individual bricks or groups of bricks might be indented or bowed inwards by more than 3mm. Alternatively, individual bricks or groups of bricks might project or bow out by more than 3mm (see Fig 128).

One of the key elements of successful plumbing up of brickwork is to spend plenty of time getting the first three courses of any new corner or wall end absolutely spot on for plumb. The positioning of the bricks in relation to the level is best examined when down on one knee. Any errors must be corrected as close to perfection as the materials will allow. Plumbing any subsequent courses is thus made easier, since the level can be 'clamped' with the foot against the first three courses and every new course above simply adjusted until it aligns with the level. In this way, the first three courses become a 'datum' for plumbing and this helps eliminate unacceptable dips, bumps and hollows in the vertical alignment.

A spirit level's bubble will indicate whether the surface plane is generally vertical (or horizontal), but a spirit level is more than just an air bubble inside a tube of liquid; it is also an invaluable straight-edge for measuring the line, flatness and

Fig. 129 It is important to spend enough time getting the first three courses spot on for plumb.

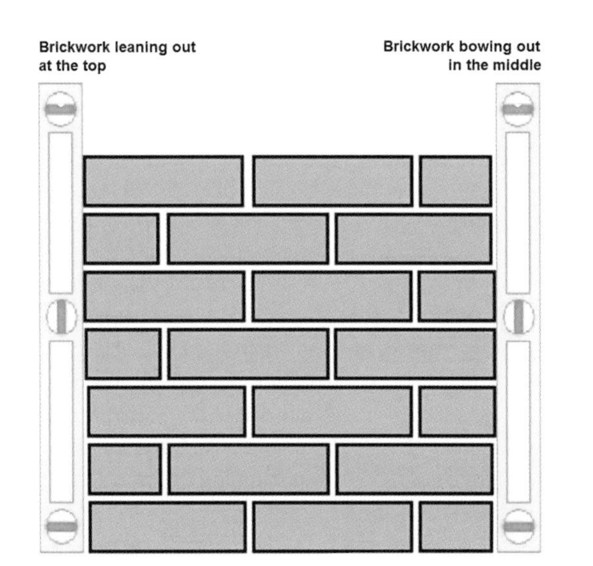

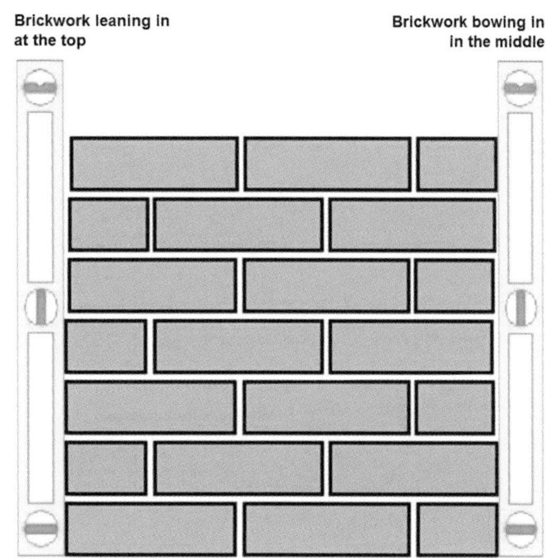

Fig. 128 Common defects when plumbing brickwork.

LEFT: **Fig. 130 Aligning the quoin brick to plumb.**

BELOW: **Fig. 131 Transferring wall lines down to foundation level.**

Setting-out line for position of wall
Spirit level
Timber profiles
Trowel marks plumbed down from setting-out lines to mark position of wall at foundation level
Thin mortar screed

face plane of brickwork. Both features together provide a bricklayer with vital information about how bricks are positioned and the extent and direction of any adjustment that may be needed. Tell-tale gaps between the brick face and the edge of the level offer useful clues. A tapering gap (*see* Fig 130) tells the bricklayer that the quoin brick needs adjustment by tapping the top of the brick at the front edge in order to bring the top of the face into vertical alignment with the level. When bricks require adjusting sideways to achieve plumb, the face of the brick should not be tapped with the edge of the trowel blade as this can damage or mark the face!

BUILDING CORNERS

The skills of gauging, levelling and plumbing brick-work need to be brought together for the purposes of building corners or 'quoins', and exactly the same principles apply to constructing the stopped-ends of wall. Accuracy for gauge, level and plumb is absolutely vital when constructing the corners or ends of walling, as any mistakes or inaccuracies will be transferred to, and replicated in, the middle section of the wall when it is run-in in between.

Setting-Out for Corners at Foundation Level

Using a spirit level, transfer the wall line vertically down to foundation level from the setting-out string-lines. Marks can be made with a trowel point in a very thin mortar screed (approximately 3mm thick) spread with the back of the trowel blade on top of the foundation.

Using the spirit level as a straight-edge and with the point of the trowel, join up the four marks to form a complete right-angle in the mortar screed. Check the right-angle for accuracy with a builder's square.

Any inaccuracy at this point will be as a result of shortcomings in the positioning of the setting-out lines and/or in the process of plumbing down from them. It will need to be investigated and corrected before proceeding. Once setting-out at foundation level is complete, the corner can be constructed (*see* Fig 132).

Building the First Course

Having set-out, always start every course with the quoin (corner) brick, which is laid to gauge, level and plumb (*see* Fig 132,A). Assuming the edges and faces of the brick to be reasonably square, plumbing the stretcher face of the quoin brick is sufficient to provide the level across the width of the brick without actually levelling it! If the bedding face and stretcher face are slightly out of square with one another, the deviation will be small and not show too much in the short bed joint along the header face. Instead of plumbing the header face of the end or quoin brick, always level along its length and then plumb up by simply adjusting the brick until the header face, at least in part, touches the level. This method is adopted because, if the bedding face and header face of the brick are slightly out of square with one another, plumbing the header face would cause a large deviation in the long bed joint along the stretcher face.

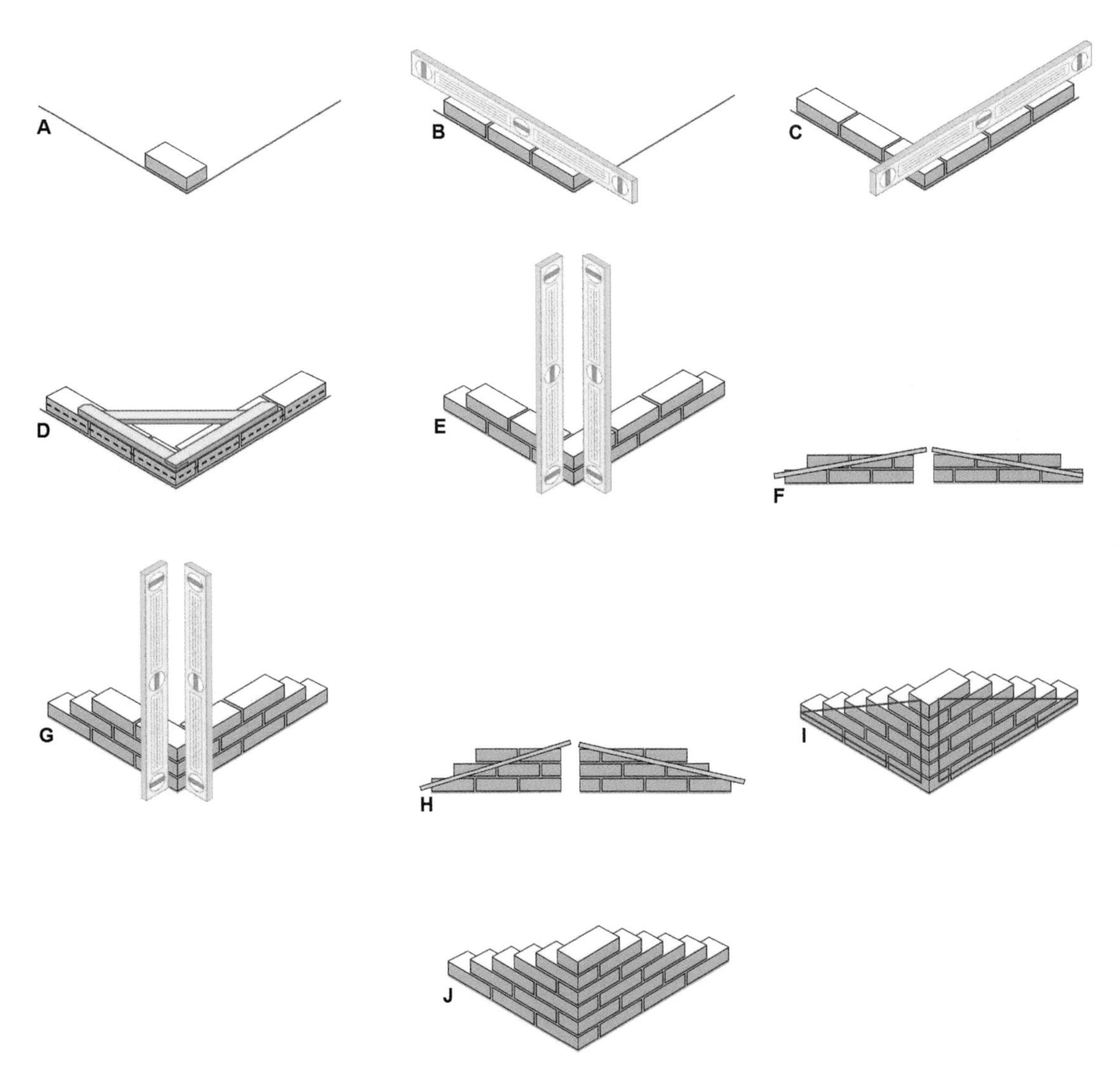

Fig. 132 The stages of building a corner.

When laying end or quoin bricks, the general rule is to plumb up the stretcher face and accept the resultant level across the width. Then the brick is levelled along its length and the resultant plumb up the header face accepted, subject to making sure the header face touches the level for plumb. Where bricks are reasonably square, this rule provides the best balance between plumb and level at a corner or end of a wall and allows for small variances in brick shape/dimensions. The only exception to this rule relates to the very first quoin brick in the first course in a new wall, which is often levelled across length and width due to the practical difficulty in plumbing up the stretcher face of one brick (only a height of 65mm), since trying to hold the level against one brick often disturbs it!

Next, construct the 'arms' of the corner (maximum of three or four stretchers long) by laying bricks, to the setting-out line, outwards from the quoin brick then levelling them along their length (*see* Fig 132, B and C). As this is the very first course of the wall, the bricks should be checked for level across their width to ensure that they are not 'tipping' over. When laying straight out from the quoin brick, care must be taken to ensure that cross-joints are of the correct thickness; if necessary, a measuring tape should be used to check the lengths of the arms. The figures in this example will be 665mm (three bricks long including 10mm joints) and 778mm (three and a half bricks long including joints).

Once the first course has been completed, it must be checked for square with a builder's square, and the spirit level should be used as a straight-edge to get both sides in line and straight (*see* Fig 132, D).

Building the Second Course

Begin the second course with the quoin brick, which is laid to gauge, level and plumbed on both faces (*see* Fig 132, E). Construct the rest of the second course and check for level along the length of each arm.

Having completed the second course, the bricks in subsequent courses are not levelled across their width but the face plane of both arms must be ranged in from the plumbing point (which is known to be vertical) down to the first course (which is known to be square and straight). Ranging in is another example of lining in brickwork between two known reference points and is achieved by using the spirit level as a straight-edge and ensuring that the face of each brick fully touches the level throughout its length and height (*see* Fig 132, F). Ranging in the face plane of every course will identify any bricks that are sticking in or out of alignment and needing widthways adjustment. It will also identify any bricks that need their level adjusting across the width; the face of such bricks will appear 'twisted' against the edge of the level by displaying tapering 'V'-shaped gaps. (It is a similar problem to that encountered when plumbing brickwork; *see* Fig 130.)

Some bricklayers will leave ranging in until after the corner has been built rather than ranging in every course individually. This is, however, bad practice as bricks that are 'twisted' in terms of face-plane alignment cannot be altered once more bricks have been laid on top of them. It can be helpful to think of ranging in as the creation of a triangle in perfect vertical alignment, where the base is the straight line of the first course, the upright side is the plumbing point, and the hypotenuse (longest side) is the ranging line between the two. Such a concept can only become a practical reality by ranging in one course at a time.

Building the Third and Remaining Courses

The third course is begun, again, with the quoin brick, which is laid to gauge, level and plumbed on both faces. Course three (and the remaining four courses) are constructed in exactly the same way as the second course, ensuring that the face plane of every new course is ranged in to the first course (*see* Fig 132, G and H) until the top of the corner is reached. The number of courses that a corner can be built is always equal to the number of bricks in the first course – in this case, seven.

Using this method of construction will result in a corner that is square, level, plumb. In addition, the faces of all courses will be in line, with no deviation in the face plane (*see* Fig 132, I and J).

PRINCIPLES OF RUNNING-IN BETWEEN CORNERS

The method of running-in is dictated by the length

Fig. 133 Laying bricks to a string-line – 'feeling' with the trowel blade whether the brick is flush with the course below.

of the wall. On short walls or piers, running-in of each course can be achieved using a spirit level. The end or corner brick ('quoin brick') is first laid to gauge, plumb and level. The end or corner brick can then be laid level and plumb at the other end, with level transferred from the first brick by way of a spirit level. Level is achieved for the brickwork in between simply by using the spirit level as a straight-edge and tapping the bricks down until the ends of the level touch the bricks at each end. Achieving the correct face line is done in the same way, but using the level on the face of each course between the two plumbing points and adjusting the face of each brick until it touches the level at its top edge and its bottom edge is flush with the course below.

Where the length of the wall exceeds the length of the level, both ends of the wall need to be built independently, with horizontal level from one end to the other being achieved using a spirit level and straight-edge. Where walls are of such a length that the use of the straight-edge becomes impractical for transferring levels, then a datum level must be transferred to each end or corner of the wall in order that the ends can be built independently but still to the same level. This is achieved by applying the methods used in Chapter 6.

For all walling that exceeds the length of a spirit level, a string-line must be used for running-in, to ensure accurate and straight alignment of brickwork in between the two ends. Bricks are laid so that the top edge almost, but not quite, touches the line, to avoid pushing the line out, and the bottom edge is flush with the course below. Pushing the line out will result in the brickwork running out of line and a bow forming in the face of the wall.

Running-in, particularly to a line, can be thought of as 'placing the face' of the brick between two upper and lower fixed points – the string-line and the top of the course below. When laying bricks and scraping off mortar that squeezes from the bed joint, it is possible to 'feel' with the trowel blade whether the bottom edge of the brick is flush with the course below. If the bed joints are spread too thickly, time and effort will be wasted in having to tap bricks down to the line. The ideal is for the mortar to be spread sufficiently thickly so that the brick can be pressed by hand to the line without any need for tapping, yet still achieving a full bed joint. As before, adjustments should not be made to the brick by tapping the face with the edge of the trowel blade; this can mark and damage the decorative face side.

Running-In the First Course

When running-in the very first course of brickwork, there is no lower reference point with which the bottom edge of the brick can be 'flushed up'. This means that it is quite easy for bricks in the first course to start 'tipping' forwards or backwards. The bricklayer should endeavour to cultivate a good eye for keeping the faces of the bricks in the first course aligned vertically, but it is a good idea to check each one with a boat level across its width.

Attaching String-Lines

When running-in brickwork to a string-line, the line is attached at each end/corner, ensuring that it is aligned with the top edge/arris of the course to be run-in. There are two basic methods of attaching string-lines: using a pair of metal pins, or using 'L'-shaped corner blocks.

When using metal pins, the ends of the pins need to be pushed into the joints of the brickwork, which causes damage to the joint that has to be made good later. Clumsy use of pins can also damage and disturb the brickwork itself. Pins have a further disadvantage in that it sometimes proves difficult to accurately align the line with the top arris of the bricks; this is made worse by their habit of working loose.

The second method uses 'L'-shaped corner blocks (Fig 135), which can be created from offcuts of timber or bought in pairs made of plastic. They are held in place by the tension of the line and have the distinct advantage of not damaging the mortar joints. However, they can only be used on external corners, which means that the use of pins is sometimes unavoidable. There must be enough tension in the line to ensure that the corner blocks stay attached and the line stays straight, but not excessively to the point that it disturbs the brickwork to which the blocks are attached. Most bricklayers avoid using pins unless they have to; the most common use of line pins is to have something to wrap the string-line around when it is not in use.

Attaching a string-line to a corner block is a four-step process (see Fig 136):

1. Pull the line straight through the front of the slot in the corner block and out through the back.

Fig. 134 Lines and pins in use.

Fig. 135 Lines and corner blocks in use.

2. Bring the string-line back round the side of the block and pull it through the front of the slot again.
3. Pull the string-line tight and repeat this stage round the other side of the corner block.
4. Pull the string-line tight to complete the process.

The string-line is usually attached to the second corner block whilst the first block is held at the other end of the wall, kept in place by a dry brick or held by a colleague. This enables the bricklayer to gauge where to position the second block on the line, so that there is sufficient tension in the line for both line and corner blocks to be attached to the wall and be self-supporting.

It is sometimes difficult to ensure that the string-line is in perfect alignment with the arris of the course to be lined in. This is a common problem when using pins as they stick out and can be easily disturbed or, as is often the case, simply work loose. Accordingly, a bricklayer will often set the line slightly higher than necessary and then 'sprag' the line down to the top arris of the course to be run-in by using a dry brick. This ensures proper alignment of the string-line and also increases the

Fig. 136 Method of attaching a string-line to a corner block.

Fig. 137 'Spragging' the line down with a dry brick.

tension in the line slightly, making the line less likely to sag. Too much tension in the line can cause brickwork in the corner to be disturbed, particularly at the top of the corner where there are fewer bricks and less mass or 'back weight' to hold them in place.

Tingle Plates

There is always the chance that lines will sag in the middle, particularly where very long walls are being run-in. If not corrected, this will result in the brickwork running out of level as it is laid to the sagging line. It is good practice, always, to check the tension in the line every few courses and re-tighten the line if required. On very long walls, the use of a 'tingle plate' is recommended. On every new course, a single brick is first laid in the middle of the wall and a tingle plate is placed on this, in order to support the line and prevent sagging. Usually home-made rather than bought, the tingle plate is a rectangular piece of thin steel which has two notches cut into one edge so that it forms three 'fingers'. The string-line is threaded over the middle finger and under the two outer fingers – never the other way round – both to support the string-line and help to maintain its correct position in line with the top edge of the brickwork being run-in (*see* Figs 138 and 139). A dry brick is usually placed on top to ensure

Fig. 138 A tingle plate.

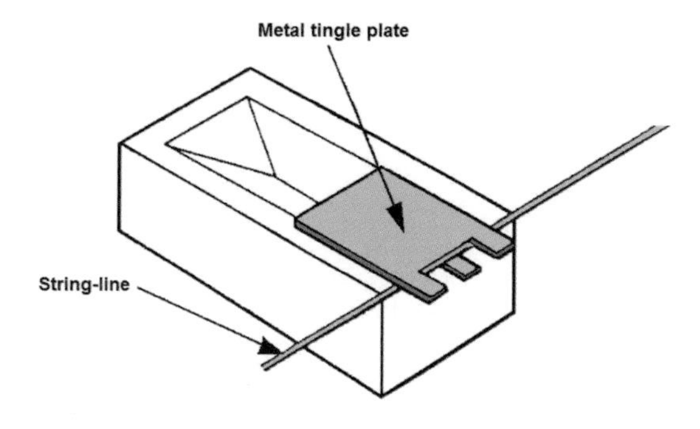

Fig. 139 Tingle plate in use to support a string-line.

Fig. 140 Furrowing under the line.

that the tingle plate stays in position; this is omitted from the illustration, for the purposes of clarity.

When making a tingle plate, it is important to file off any sharp edges between the tingle plate fingers that could cut or damage the string-line.

Disturbing the Line

In order to maintain accuracy when running-in to a string-line, it is important that the line remains as undisturbed as possible. Bricks should be laid as close to the line as possible, but it is too easy to catch the line with the trowel blade and move it out of position, which will affect the finished brickwork. When spreading bed joints, a conscious effort should be made to lift the trowel over the top and well clear of the line. The action of furrowing the mortar bed is the most common cause of the line being disturbed as the line can get pressed down with each downward part of the furrowing action. To avoid this, turn the trowel horizontally through 90 degrees and furrow side to side under the line. This method of furrowing (*see* Fig 140) is achieved by tilting the trowel blade in the direction of furrowing to push and pull the mortar in both directions along the top of the wall.

> **Disturbing the Line**
>
> When two bricklayers are working on a long wall and running-in to the same line, it is important to be aware that one bricklayer constantly 'twanging' the line can be an enormous source of irritation to the second bricklayer working from the other end!

Laying the Last Brick to Line

The last brick in a course of brickwork can be awkward to lay, whether it is to a line or not.

Fig. 141 Laying the last brick to a line.

Cross-joints should be applied to both ends of the last brick but the brick should not be pushed straight down vertically into the gap, as this can cause the cross-joint mortar to be scraped off by the bricks either side. Instead, the brick is reversed in and down at an angle, the cross-joint is squeezed up and then the front end is lowered down (*see* Fig 141).

FACE-PLANE DEVIATION

Following running-in, be it to a spirit level on a small job or a string-line on a large job, it is necessary to check the deviation of the face plane of the wall. This involves placing a straight-edge on the face of the wall and across both diagonals in order to check the flatness of the finished brickwork (*see* Fig 142). On short walls this can often be achieved with the edge of the spirit level.

Checking the face-plane deviation is not only a measure of how well the brickwork has been run-in but also a measure of how well the two ends or corners of the wall were originally constructed. Any error in the construction of the ends or corners will be transferred to, and replicated in, the middle section of the wall when it is run-in! The accepted tolerance for face-plane deviation is no more than 5mm in any 3m length.

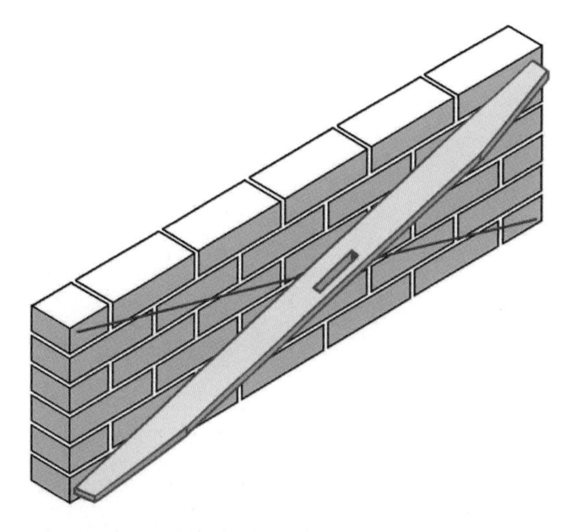

Fig. 142 Using a straight-edge to check face-plane deviation.

Technical Correctness versus Aesthetics

Part of a bricklayer's job is to create the best possible balance between plumb, level, line and face-plane deviation and aesthetics. It is not always possible to adhere to industrial tolerances, due to the limitations of the materials being used. Creativity is sometimes needed because bricks – man-made components from a naturally occurring substance – are never perfect. They can vary greatly in shape, size and squareness, and the bricklayer must allow for this in his or her work by doing whatever is necessary to strike a balance between what is technically correct and what looks right and pleasing to the eye. Sometimes, technical accuracy may have to be sacrificed in favour of aesthetics, with the bricklayer altering his or her approach to suit the nature of the bricks being laid. On those occasions, bricklaying becomes less of an exact science and more of an art!

CONSTRUCTING BLOCK WALLS

Principles

Despite their size (six times bigger than bricks), the craft operations in erecting walls and corners from blocks are much the same as for bricks, with a couple of exceptions. The vertical height and face area of a typical block greatly exceeds its width and bedding face area, so, when laying blocks for the first course of a corner, they should never be levelled across their width. Instead, they should only be levelled along their length and plumbed vertically up their face. The latter, assuming the faces of the blocks are square (which they usually are), will provide more than adequate provision for level across their width. This principle applies to a lesser degree where much wider blocks are used, but reference is made here to blocks that are 100mm wide.

When running-in the first course of a block wall, the large height of blocks and the lack of a lower reference point to which to align the bottom edge make it quite easy for blocks to end up leaning over one way or the other. To avoid this it is common either to line in the bottom edge of the first course

Fig. 143 Building block walls and lining in the first course with a straight-edge.

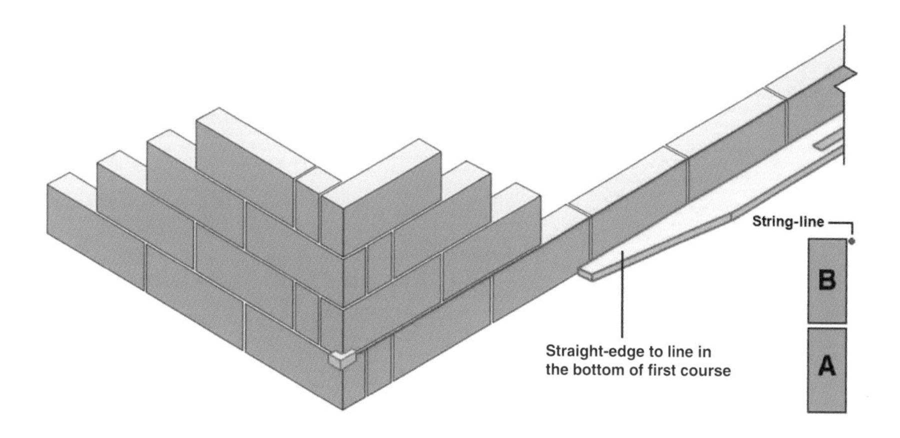

with a long straight-edge between the corners (see Fig 143), or, if the wall is too long, to attach a second string-line as close to the bottom edge as possible in order to create a lower reference point to which to align the faces of the blocks.

A good flat face-plane block for blockwork walling is vital – one block that is out of vertical alignment is much more noticeable than one misaligned brick, because of its size. That said, an accurate face plane still depends upon adopting exactly the same principles as when laying bricks to a line (see Fig 143). Block B must be laid with its top edge parallel to the line (as close as possible to the line without touching it) and its lower edge flush with the top edge of block A.

Laying Blocks

Blocks should not be tapped excessively, or at all if possible, to align them, as this tends to make them lose adhesion with the mortar bed. This is particularly the case with aerated concrete blocks, which have a higher suction rate and tend to dry out the mortar bed quite quickly. Instead, try to squeeze the blocks into position, which means avoiding spreading joints that are too thick. A common mistake made by many people who are new to bricklaying is to assume that, because a block is big, the mortar joints should also be big. This is simply not the case. When laying blocks, bed joints and cross-joints are 10mm, as they are for brickwork. The size of blocks has been designed to complement bricks, with one course of blocks being the same height/gauge as three courses of brickwork and one block being the same length as two bricks with a 10mm cross-joint in between (Fig 144).

When laying aerated concrete blocks for fair-faced work, be careful not to damage arrises during handling and laying. Aerated blocks break easily if dropped and carelessly trimming off mortar with the trowel when spreading bed joints can easily cut the edge of a block. Similarly, the block face is easily damaged when adjusting the block to position by careless use of the edge of the trowel blade.

Applying Cross-Joints

When applying cross-joints to blocks, which need three times as much mortar as a brick, it helps adhesion if the mortar can have something to stick to. Smearing the end face of the block with a little mortar makes it easier to apply the cross-joint, without the mortar falling off before the block is laid. Remember that the cross-joint must be full!

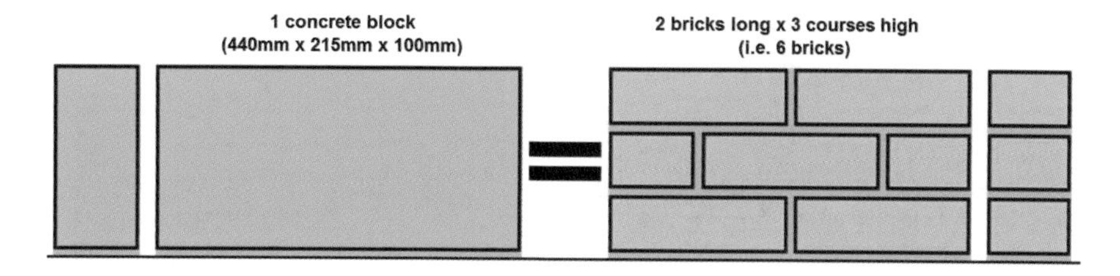

Fig. 144 Size comparison between bricks and 100mm blocks.

Constructional Height Limits of Blockwork

Blockwork that is being built as part of a cavity wall is supported by the wall ties that form an integral part of the cavity-wall structure, so there are no significant practical limits on how high such blockwork can be built in one day. Free-standing block walls or partitions, however, can begin to buckle under their own weight if built too high. In addition, very dense, heavy blocks with low water absorption can begin to sink and squeeze out the lower bed joints if they are built too high in one go. To avoid such issues, the height of new blockwork in free-standing walls should be limited to a maximum of six courses.

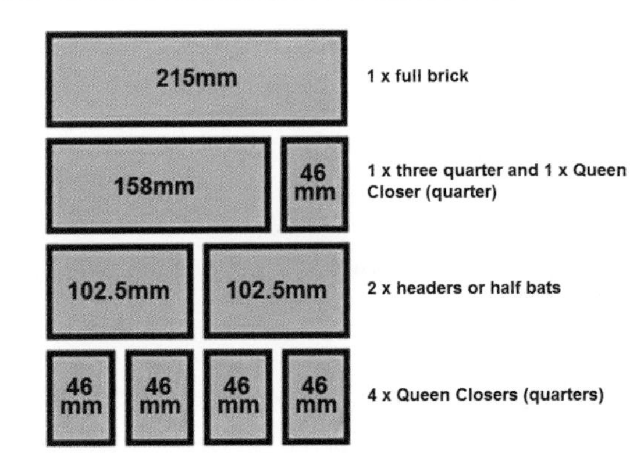

Fig. 146 Co-ordinating sizes of bats and closers with an allowance of 10mm for mortar joints.

CUTTING BRICKS AND BLOCKS

Depending on their finished size, cut bricks are referred to as 'bats' or closers'. In the same way that the dimensions of full bricks have been designed so that they coordinate together, with an in-built allowance for a 10mm mortar joint, standard bats and closers such as quarters, halves and three-quarters are cut to specific dimensions that also make an allowance for a 10mm mortar joint. For example, two Queen Closers at 46mm each, plus 10mm for a joint, will be equal to half a brick – well, almost, since a half-brick is actually 102.5mm (the same as the header face of a brick), but it would take an unachievable level of accuracy to cut a brick by hand to a tolerance of 0.25mm. For the same reason, cutting a half to 102mm is considered acceptable, with the resultant 'missing' 0.5mm easily being accommodated in the adjacent cross-joint.

Fig. 145 Half bat, three-quarter bat, Queen Closer and full brick.

Fair-Faced Cutting

When carrying out the fair-faced cutting of bricks (that is to say, making cuts that will be seen), the cuts should be correctly sized. Where they repeat, such as in a broken bond, they must also all be of equal size; failure to achieve this will result in deviation in the verticality of cross-joints and odd-sized cross-joints. The actual line of the cut should be neat and straight, with no spalling or damage to the brick face. The use of sharp cutting tools is a key contributor to this last point.

Standard cuts for quarter bats, half bats and three-quarter bats can be easily and repetitively measured and marked out using a 'closer gauge', sometimes known as a 'bat gauge'. This is usually home-made from off-cuts of timber. They are not that widely used and most bricklayers, when measuring and marking half bats (by far the most common cut brick), will simply use the width of another brick as a ready-made 'gauge'.

For neat, accurate cuts on good-quality facing work, a sharp bolster chisel and lump hammer are used. The brick to be cut should have a good

Fig. 147 Wooden closer gauge, also known as a bat gauge.

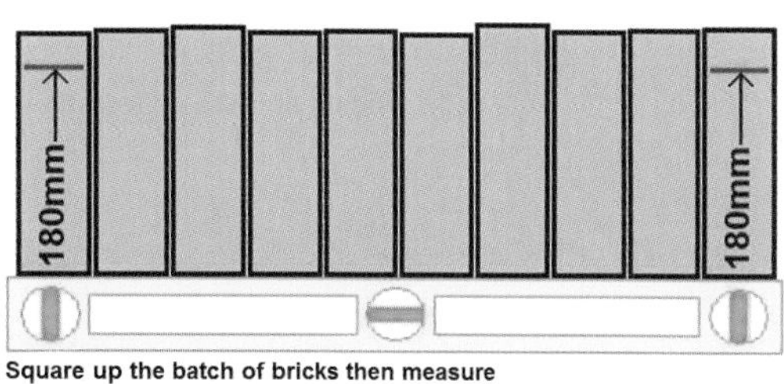

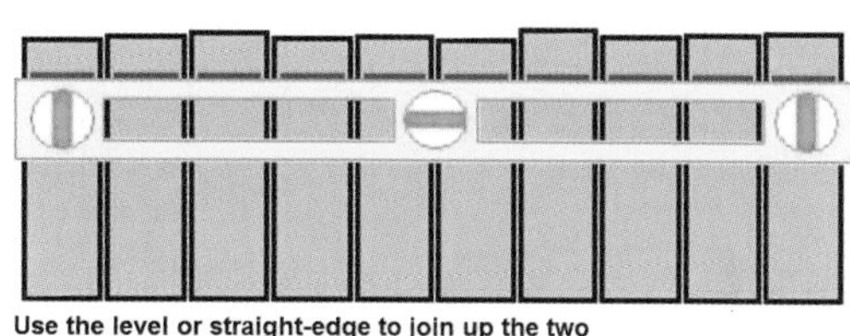

Fig. 148 Marking a batch of bricks for cutting.

face and appear sound, with no obvious cracks, as defective bricks are more likely to shatter or break in the wrong place.

Measure the bat or closer required – for standard cuts, this can be done with a closer gauge – then mark the cutting line with a pencil or the corner of the bolster blade. Non-standard cuts, for use in a broken bond for example, will need to be measured with a tape and marked to the required size, with allowance for cross-joints. To save time and to ensure that all the cuts are of equal length it makes sense to measure and mark several bricks at the same time (see Fig 148). Start by lining up a number of bricks (in this case ten) face side up and square up the ends with a spirit level or straight-edge. From the squared-up edge, measure and mark with a pencil the two end bricks to the desired length, in this case 180mm. Line up the spirit level or straight-edge with the pencil marks on each end brick and mark the remaining bricks in between to the same length. This method not only saves time but also eliminates any error that can result from the common variations in brick length.

Place the brick marked for cutting on a piece of 'softing'; this will take the 'jar' out of cutting and reduce the risk of the brick breaking in the wrong place, which can happen if it is cut on a hard surface such as concrete. Ideal materials for softing include rubber matting and off-cuts of carpet or old carpet tiles. If none of these materials is available, spread some soft sand about 25mm thick on top of a hard surface.

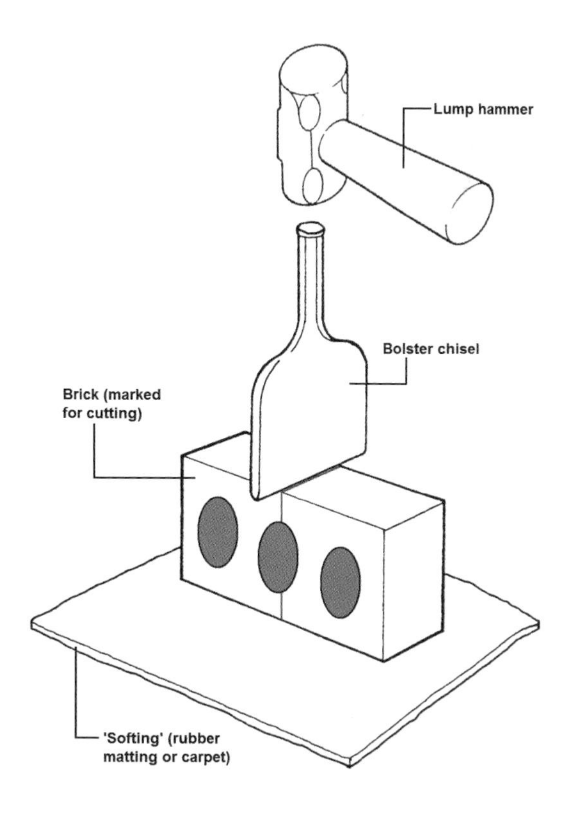

Fig. 149 Hand-cutting bricks on a piece of 'softing'.

When cutting, gloves and eye protection in the form of safety glasses or goggles must be worn.

Place the brick face side up on the softing and strike the bolster once on the cutting mark with the lump hammer. The blow should be hard enough to mark the face but not hard enough to break the brick in one go.

Turn the brick through 180 degrees, with its common face uppermost, and repeat the process on the opposite side of the brick. Finally, turn the brick through 90 degrees on to its bedding face and cut again between the two original cuts. If the brick does not break at this point, which is unlikely, turn it over through 180 degrees on to its other bedding face and cut again. It is fair to say that the amount of power applied with the hammer needs to vary depending on the hardness of the brick being cut but at all times it is advisable to think in terms of 'encouraging' the brick to break, rather than forcing it. Too much effort will result in shattered bricks and excessive waste.

Any final trimming of the cut bat/closer that may be required can be done using a brick hammer or a scutch hammer fitted with either a blade or comb. Finally, unwanted brick debris should be cleared away as soon as possible to ensure the work area remains tidy.

Cutting Bricks at an Angle

Not all cutting is done at square angles. For some bonding arrangements it is necessary to cut a brick at an angle on plan and for certain types of decorative work there is a need to cut bricks at an angle on face. The general methods used are much the same as those used for square-angled cutting, but a particular approach in terms of the order of things must be adopted. When cutting bricks at an angle on face, the brick should not be cut in one go. The waste must be cut off first with a square cut, otherwise the acute angle will virtually always break off. Frogged bricks should be cut so that the acute angle is part of the 'meat' of the brick and not the frog, which would be so fragile that it would always break off regardless of how well the brick is cut.

Rough-Cutting Bricks

For the rough-cutting of bricks for unseen work, it is possible to use the cutting edge of the brick trowel, simply holding the brick in the laying hand, estimating by eye the length of the cut brick required and giving a sharp blow with the cutting edge of the trowel to remove the excess. However, this method is not recommended and, over time, it will shorten the life of the trowel, particularly when it used on harder engineering bricks. It is much better to use a brick hammer that has been designed for rough-cutting and can provide a greater degree of ease and accuracy.

When rough-cutting bricks, try to cultivate a good eye for brick length. It can help to offer up the brick to be cut to the gap where it is intended to fit, and mark it with the blade of the brick hammer,

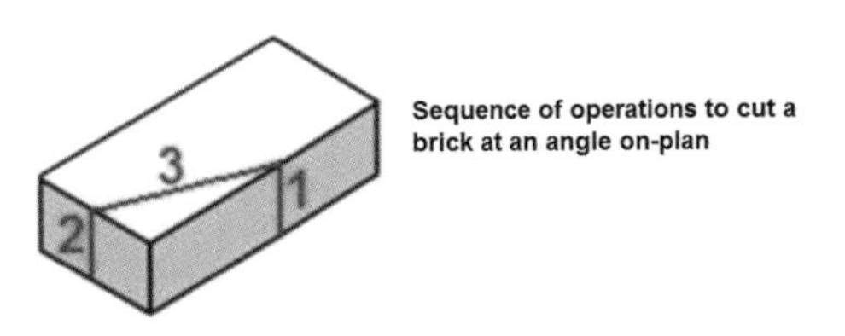

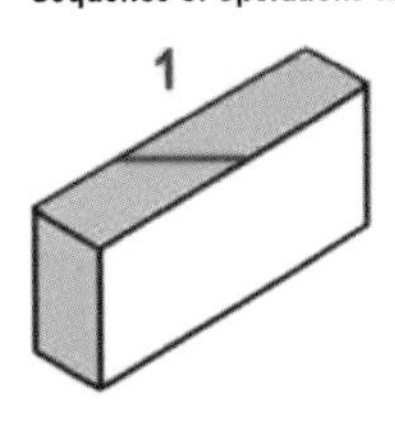

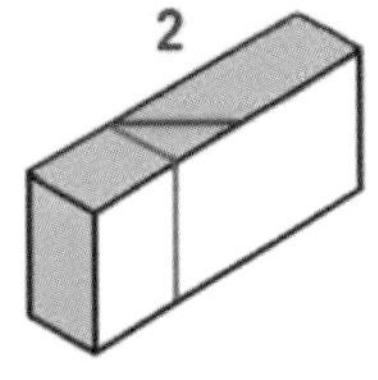

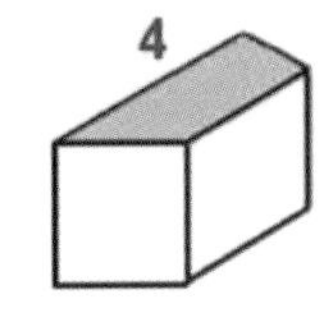

Fig. 150 When cutting bricks at an angle, each step must be carried out in the correct order.

Fig. 151 Petrol-driven STIHL saw.

not forgetting to make an allowance for cross-joints at both ends.

Cutting Blocks

Concrete blocks are generally cut using the same method as for bricks, with the bolster chisel and lump hammer. It is usual, however, to have to cut the block on all four sides before hitting it hard enough on its bedding face to break it. Aerated concrete blocks can also be cut with a bolster chisel and hammer but the material, which is less dense, tends to absorb more of the impact shock and does not crack as easily or as accurately as a concrete block or a brick. The simplest and most accurate method of cutting aerated blocks is to use a masonry saw or even an old joinery panel saw. The material is soft and offers so little resistance to cutting that an aerated block is easier to cut than most timbers.

Mechanical Cutting

On a building site bricklayers will often use an electric angle grinder with a diamond steel blade or stone-cutting disc to cut both concrete blocks and bricks. Petrol-driven versions are available, and these are capable of cutting practically anything. Regulations do not permit persons below the age of eighteen to use these machines and all users should wear steel toe-capped boots, goggles and ear defenders.

PROTECTING NEW BRICKWORK AND BLOCKWORK

At the end of every working day, newly laid brickwork or blockwork should be protected from the actions of rain, frost or snow. Rain will wash unset mortar from the joints, leaving stains down the face of the wall, while icy conditions will cause the water in the unset mortar joints to freeze and then expand, causing irreparable physical damage to the new brickwork, particularly if the top of the wall is allowed to become saturated. Allowing the top of the wall to become saturated can also increase the likelihood of free lime leaching from mortar joints (causing the effect known as lime staining). It may also lead to unsightly efflorescence, which occurs when soluble salts from the brickwork remain on the face of the wall following drying out. The action of the sun in hot weather cannot be ignored either. High temperatures in summer can cause mortar joints to dry out too quickly, before the cement has set. This problem is exacerbated by prevailing winds, which draw moisture from the brickwork.

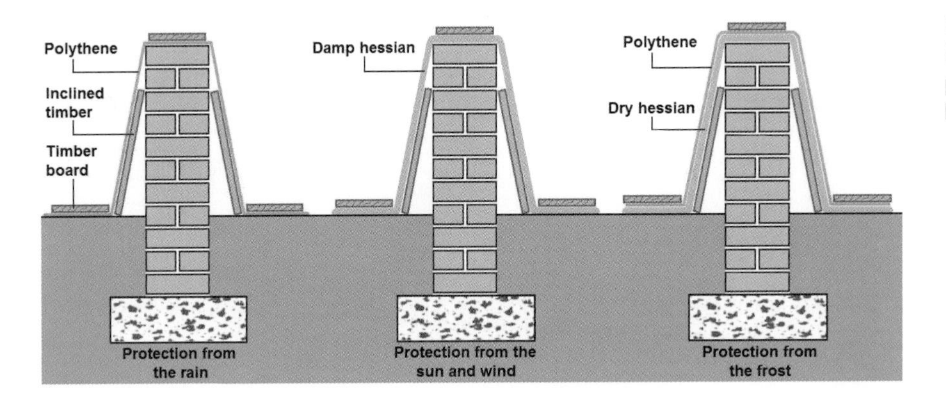

Fig. 152 Methods of protecting new brickwork.

The prevailing weather conditions will dictate the method used to protect new external walling; there are various approaches that can be taken.

During periods of wet weather, the brickwork should be protected with polythene sheeting, weighted down on top of the wall and at ground level with timber boards or scaffold battens. In addition, inclined timbers are leant against the wall every metre or so, in order to hold the polythene away from the faces of the wall. This creates an airspace that allows the wall to breathe, creates some ventilation and avoids condensation forming. Condensation can be just as injurious to the wall as rain. The need to create some ventilation in this way is common to all the protection methods, regardless of weather conditions. During warmer months, the action of rain is less of an issue, but it is worthwhile protecting against the detrimental effects of sun and wind by draping the wall with damp hessian. Ensure the hessian is not too wet, as this can result in staining from the mortar joints. The main problems during winter are frost and rain, meaning that the wall has to be kept insulated and dry. This is achieved by using dry hessian for insulation, with polythene sheeting placed over it. The polythene keeps the hessian dry so it retains its insulating qualities, and ensures that the wall is kept dry at the same time.

CALCULATING QUANTITIES OF MATERIALS IN BRICK AND BLOCK WALLING

Bricks and Blocks

When calculating the quantity of bricks or blocks in a wall, a mathematical constant is used in each case. For calculating bricks, this is 60 bricks in every square metre of half-brick thick walling. (For one-brick walling, therefore, the calculation is 120 in every square metre.) For calculating blocks, the constant is 10 in every square metre (on the basis that a block is six times the face area of a brick). Over and above these figures, it is necessary to add a percentage for cutting and wastage – a maximum allowance of 10 per cent is usually sufficient.

Mortar for Laying Bricks and Blocks

The most basic estimate of materials for mortar states that one tonne of sand is sufficient to lay 1000 bricks. Taking this a stage further, a quick and convenient method for calculating materials for mortar involves working on the basis of the amount of finished mortar that is required to lay one brick. This amount will vary according to the type of brick. For example a pressed brick with a 'frog' in it requires more mortar than a solid brick with no perforations or indentations. However, as a general rule of thumb, it can be taken that 1kg of mortar is sufficient to lay one brick. Where 100mm blocks are concerned, the amount of mortar required per block is 2.25kg.

Beyond this, and the number of bricks or blocks in the wall, the only additional information required in order to calculate the quantity of each component part of the mix is the ratio of the mortar mix proportions. There is no need to add an allowance for waste when calculating mortar materials as this has already been accounted for when adding an allowance on top of the brick or block quantity.

Calculating Materials by Weight

Calculating quantities of materials for mortar by weight is convenient, as that is how the materials are sold, and it also eliminates the need to take account of volume shrinkage on mixing! When working out quantities of bagged materials, the number must always be rounded up to the next full bag. Rounding down will lead to insufficient quantities. Caution should also be exercised when buying large quantities of sand in bags – it is likely to be much more economic to buy it loose by the tonne.

Simple Worked Example for Brick Walling

Project: a one-brick thick wall 12m long and 2.6m high (measured from the top of its foundation), constructed using a mortar mix of 1:2:4 (cement:lime:sand).

Number of bricks = 12m × 2.6 × (60 per square metre × 2) = 3744 bricks
Add an amount for waste (typically 10 per cent): 3744 + 10 per cent = 4118 bricks

Quantity of mortar = 4118 × 1kg/brick = 4118kg in a ratio of 1:2:4 (total of 7 component parts)
Cement: (4118kg ÷ 7 parts) × 1 = 588kg ÷ 25kg/bag = 23.53 = 24 bags
Lime: (4118kg ÷ 7 parts) × 2 = 1177kg ÷ 25kg/bag = 47.08 = 48 bags
Sand: (4118kg ÷ 7 parts) × 4 = 2353kg ÷ 25kg/bag = 94.12 = 95 bags

Simple Worked Example for Block Walling

Project: a 100mm-thick block wall 12m long and 2.6m high (measured from the top of its foundation), constructed using a mortar mix of 1:2:4 (cement:lime:sand).

Number of blocks = 12m × 2.6 × 10 per square metre = 312 blocks
Add an amount for waste (typically 10 per cent): 312 + 10 per cent = 343 blocks

Quantity of mortar = 343 × 2.25kg/block = 772kg in a ratio of 1:2:4 (total of 7 component parts)
Cement: (772kg ÷ 7 parts) × 1 = 110kg ÷ 25kg/bag = 4.40 = 5 bags
Lime: (772kg ÷ 7 parts) × 2 = 221kg ÷ 25kg/bag = 8.84 = 9 bags
Sand: (772kg ÷ 7 parts) × 4 = 441kg ÷ 25kg/bag = 17.64 = 18 bags

CHAPTER 10

Basic Joint Finishes

POINTING AND JOINTING

A finish is applied to brickwork mortar joints for a number of reasons. In general terms, a combination of aesthetic and weather-resistance considerations, dependent upon the location of the brickwork and/or its exposure to weather, determines the nature of the joint finish.

The main objective of applying a finish to mortar joints on external brickwork is to seal and compact its surface in order to prevent water penetration into the wall. The secondary objective is to provide a decorative finish to the brickwork. On internal brickwork, the choice of joint finish is wholly an aesthetic one. There are a variety of different joint finishes available, a number of which are more commonly used than others.

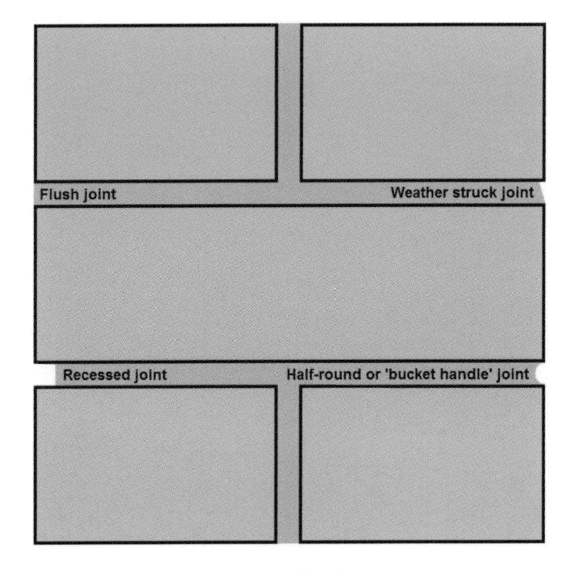

Fig. 153 Common joint finishes.

It is important to understand the difference between the terms 'pointing' and 'jointing', which are often confused and used incorrectly. 'Pointing' refers to the filling of existing joints with mortar followed by the application of a joint finish (in other words, jointing). It is generally associated with the raking out of decayed mortar joints (to a depth of 15mm) in old or existing brickwork, followed by re-pointing with new mortar and the application of a joint finish.

'Jointing' is the term generally used when the mortar joints have a joint finish applied to them as the work proceeds. It is often referred to as 'jointing-up' and is a process associated with the construction of new brickwork.

Jointing should never be considered as merely a minor detail or something to be rushed. The choice of joint finish, the skill with which the joint finish is applied by the bricklayer, and allowing enough time to joint-up all have a fundamental bearing on the durability and appearance of finished brickwork. Poor-quality jointing can make excellent brickwork look distinctly average, whereas good-quality jointing can give average brickwork a visual lift, to the point where it can look quite good!

Importance of Jointing

A measure of the importance of applying skill and care to the jointing-up process is the fact that mortar joints account for around 20 per cent of the face area of a brick wall. It is not a small element by any means.

TIMING OF JOINTING-UP

The right time to joint-up new brickwork is a key consideration. The mortar should still be soft enough to allow a jointing tool to compress it easily into full contact with the arrises of the bricks. Attempting to joint up when the mortar is too soft or wet has the effect of drawing cement paste to the surface in a watery, rivened pattern. When it is dry, this leaves the mortar joint with a porous surface and reduced weather resistance. Jointing-up too soon also causes the mortar to spread on to the brick face, leading to staining. It will also be uneven, as the overly soft mortar offers little resistance to the jointing tool, making it difficult to maintain a consistent line or depth. Jointing-up when the mortar is too dry or hard means excessive force is required to compress the joint and this may result in the surface of the joint becoming blackened by over-rubbing, and create a crumbly and porous texture.

The precise timing of jointing is affected by various factors, including the suction rate of the bricks, the existing water content of the bricks, the mortar mix design, and the temperature, humidity and prevailing wind. For example, bricks with low water absorbency or suction rate, such as Staffordshire Blue engineering bricks, can be laid first thing in the morning and will not be ready to be jointed until the late afternoon. Cold and/or wet weather will cause mortar to remain soft for longer, thus lengthening the period between laying and jointing-up. Conversely, very dry bricks with a high suction rate might need to be jointed up after laying just a few bricks. This is made worse in warm, dry weather, when the mortar can 'go off' so quickly that it prevents an adequate bond between brick and mortar before jointing-up has even been considered. To try to mitigate the effects of these extremes, bricks of low water absorbency should be kept dry, while bricks of high absorbency may be sprayed with water to make them practical to use.

The timing of jointing-up is clearly not an exact science due to the number of variable factors that affect it. The best course of action is to test areas of brickwork periodically to see if they are ready to receive a joint finish. If so, jointing-up can be carried out; if not, it is wise to wait a little longer.

Pointing Prior to Jointing-Up

When laying bricks, one objective is to ensure that the joints are 'flush from the trowel', to facilitate the jointing that will follow. This means that, when the excess mortar is trimmed off, both cross-joint and bed joint should be full and show no visible gaps that need to be filled in (pointed) before the jointing process starts. Inevitably, some pointing will be necessary prior to jointing-up new work, but fresh mortar should not be used. One solution is to set aside a trowel full of mortar on a dry brick, which will 'go off' somewhat and can then be used for any pointing that may be needed. Being slightly older, it will have a consistency that is suitable to receive a joint finish immediately after. For more on pointing and re-pointing, *see* Chapter 14.

HALF-ROUND OR 'BUCKET HANDLE' JOINT FINISH

A half-round joint is produced by 'ironing' the joints with a rounded jointer or jointing iron. The reference to 'bucket handle' comes from the historic use of galvanized metal bucket handles, which bricklayers adapted into jointing tools. Modern half-round jointing irons are commercially available but many bricklayers make use of a length of 13mm diameter mild-steel bar that is cranked to provide a handle. The diameter of the bar should ideally exceed 10mm, to avoid joints being ironed excessively deep and to maintain a consistent depth of finish.

Fig. 154 Half-round joint finish.

TOP: **Fig. 155 Correct use of a half-round jointing iron.**

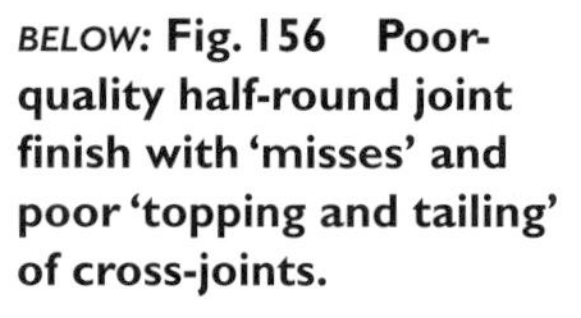

BELOW: **Fig. 156 Poor-quality half-round joint finish with 'misses' and poor 'topping and tailing' of cross-joints.**

When applying a half-round joint finish, all the vertical cross-joints should be done first. Care should be taken to keep the jointing tool straight and not to dig in to the centre of the cross-joint, as this will give the joint a concave appearance through its height. Once all the cross-joints have been done, the bed joints should be jointed. Care should be taken to ensure that all mortar squeezed out at the arrises of the bricks is carefully and neatly trimmed off with the pointing trowel or tip of the brick trowel.

Jointing the bed joints will have the effect of 'closing off' the top and bottom of all the cross-joints, making it necessary to revisit the cross-joints and, using the heel of the jointer, 'tuck in' the top and bottom of each one. This is often referred to as 'topping and tailing' the cross-joints and gives a uniform and continuous finish with the bed joints.

When ironing the joints, it is important to ensure that the jointing tool stays in contact with the arrises of the bricks above and below the bed joints and either side of the cross-joints (see Fig 155). This ensures good contact between the mortar joint and the bricks at their arrises, an even depth to the joint finish and a continuous surface to the joint, which should not suffer from 'tram-lines' or 'misses'.

The use of half-round jointing tends to be quite forgiving of minor deviations in the bricks or brick-work, since it produces a rather homogenous surface over the face of the wall where bricks and mortar joints blend in to each other. In addition, half-round jointing is simple and quick to apply and provides a good level of weather resistance. For all these reasons, it is by far the most common joint finish in use today.

RECESSED JOINT FINISH

Recessed jointing exposes every face edge of every brick in a wall, creating heavy shadow lines across the wall. As a result, it throws into very sharp relief any defects in the bricks and/or brickwork.

Fig. 157 Recessed joint finish.

Accordingly, this joint finish should be limited to situations where only the best bricks and bricklaying have been employed. The exception might be where a very 'rustic' look is desirable; indeed, it is not uncommon to see a recessed joint applied to walls built with hand-made bricks of varying shapes.

Recessing is carried out using a 'chariot' tool or, more commonly, a home-made raking tool. This is easily manufactured out of a small block of wood (50 × 30 × 20mm) and a round-headed nail that projects from the face of the wooden block by a distance equal to the required depth of the joint finish. The flat head of the nail cuts into the mortar joint and leaves the face of the joint with a flat surface. Cutting a rebate into the block of wood at the point where the nail is to be hammered in prevents the mortar that is being raked out clogging around the nail and staining the face of the brickwork. A chariot is no more effective than the home-made version; its advantage is that the nail or spike, which is inserted and locked into the tool with a turn-screw, can be adjusted for different depths. Also, a block of wood will wear out with short-term use whereas a chariot will not.

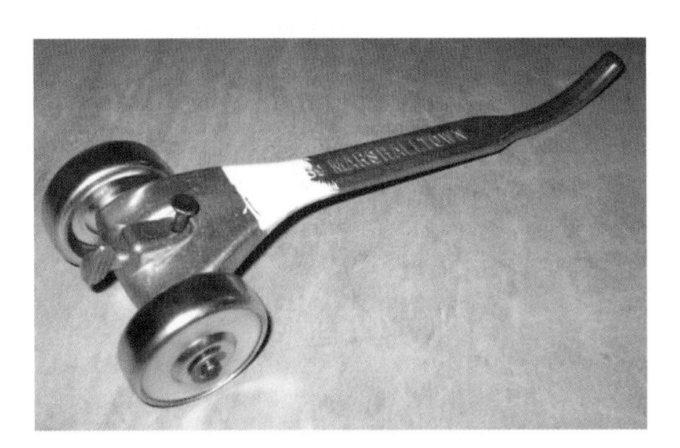

Fig. 158 'Chariot' recessed jointing tool.

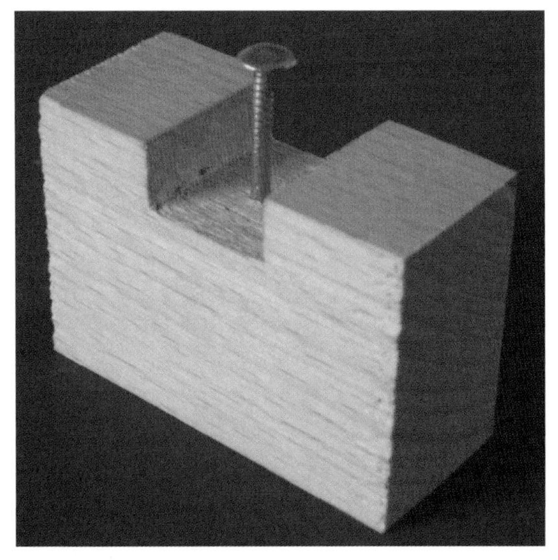

Fig. 159 Home-made recessed jointing tool.

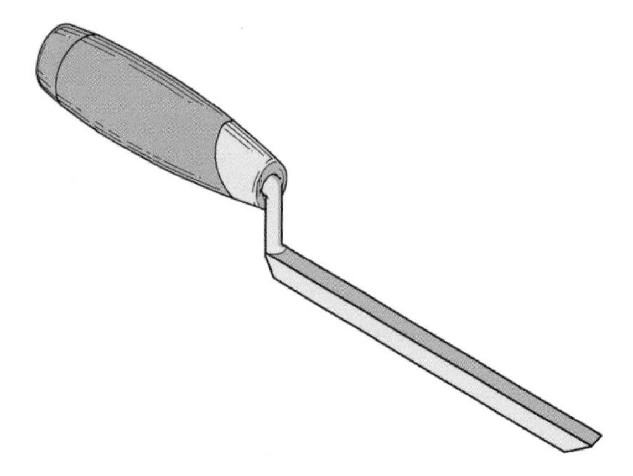

Fig. 160 Flat/recessed jointing iron.

In recessed jointing, joints are typically raked out to a depth of 10mm. The cross-joints are always raked out first, followed by the bed joints, taking care to ensure that the raking out is uniform to the full depth. There should be no mortar left clinging to the exposed brick surfaces inside the recess.

On internal work, the raked-out joint can be left 'rough' and just give a light going-over with a soft brush to remove any debris left inside the recessed joint. On external work it is necessary to perform the additional task of 'polishing' the flat surface with a steel, square-edged flat/recessed jointing iron, to increase the weather resistance of the joint. Even after this extra polishing up, this type of joint will have significantly less resistance to rain and frost than, say, a half-round joint, which is much more compressed at its outer surface. In addition, the recess of the joint offers an ideal ledge upon which rainwater can collect, resulting in eventual frost damage to the arrises of the bricks in winter, unless hard-burnt bricks of engineering quality have been used. In general, recessed jointing is most commonly applied to internal brickwork.

Fig. 161 Weather-struck joint finish.

WEATHER-STRUCK JOINT FINISH

A weather-struck joint is produced by ironing the joint with the back of a pointing trowel and, at the same time, pressing or insetting the bed joint at the top edge behind the arris of the brick above. This forms a downward, flat sloping surface that finishes at the top arris of the bricks below, allowing rainwater to run off (inspiring the name of the joint). Cross-joints are inset on one side, usually the left. Whether left or right, however, all cross-joints must be struck to the same side, otherwise there will be odd shadow-line effects on the face of the wall.

The depth of inset must be kept uniform and consistent and the thickness of the pointing trowel is generally regarded as being sufficient. Any deeper insetting tends to result in bold shadow lines that spoil the face of the finished brickwork. Again, cross-joints should be jointed first, followed by bed joints, taking care to ensure that all mortar squeezed out at the arrises of the bricks is carefully and neatly trimmed off with the pointing trowel or tip of the brick trowel. As with half-round jointing, once the bed joints have been finished, the cross-joints must be 'topped and tailed', to blend them into the bed joints.

When applying weather-struck joints, care must be taken to ensure that the full depth of bed joints and full width of cross-joints are ironed to a flat sloping finish that meets with the arrises of the adjoining bricks. If this is not done, flat spots will be apparent when trimming off the excess mortar and the joint will have to be ironed again.

An alternative to a weather-struck joint is a struck joint, where the bed joint is inset on the bottom edge. Its limitations on external works are similar to those that apply to a recessed joint.

Fig. 162 Struck joint finish.

Fig. 163 Flush joint finish.

FLUSH JOINT FINISH

As the name implies, this joint finish is flat and intended to be flush with the face of the brickwork (*see* Fig 163). On internal work the joints can simply be left 'flush from the trowel' while the bricks are being laid – this 'rough' finish can contain small gaps or misses and offers no weather protection if used externally. Ironing the joints with a tool made of steel will leave tool marks in the mortar so a wooden tool (200 × 50 × 12mm) with a rounded spatula-like end should be used to flatten the joints. Care must be taken to ensure that the tool does not cause the joints to look concave; they should be properly flat.

When applying this type of joint finish, timing is very important, as it is very easy to stain the face of the brickwork. If it is done too soon, the mortar will be too wet; too late and the finish will be crumbly.

On internal work, a joint left flush from the trowel can be given a very uniform and gap-free finish by rubbing the surface with a small piece of sack cloth ('bag rubbing'). Whilst the surface finish is even, this process does expose the aggregate in the mortar as opposed to compacting the joint, so it is not recommended for external work. On large areas the sack cloth should be turned periodically and shaken out, to avoid debris staining the face of the brickwork.

ATTENTION TO DETAIL

Poor jointing can spoil the finished appearance of even the best brickwork, and timing, care and attention to detail are key factors in the process.

Bed joints at any external angles, stopped-ends and reveals should be neat and sharp (*see* Fig 164). To achieve this, it is necessary to iron the joint away from the external angle rather than towards it, as the latter can cause the end of the joint to be rounded over or 'dragged out', particularly if

Fig. 164 Half-round jointing correctly finished at an external angle.

Fig. 165 Half-round jointing correctly finished at an internal corner.

Fig. 166 Poorly finished half-round jointing at an internal corner.

excessive pressure is applied with the jointing tool. The same principle applies to the cross-joints between bricks laid on edge, where any lack of attention to detail will be very noticeable, due to the sheer number of joints involved.

Cross-joints at internal angles must distinctly follow the bonding arrangement and the intersection of the bricks on individual and alternate courses (*see* Fig 165). Under no circumstances should the cross-joints on alternate courses 'blend together' to the point where they practically appear to be one continuous vertical joint (*see* Fig 166).

Brushing over the brickwork with a soft hand brush, to remove the last debris or crumbs of mortar not already removed by the trowel or jointing tool, is recommended. However, this should be carefully timed so as not to cause damage to the joint finish or to leave brush marks in mortar that is still too wet.

CHAPTER 11

Cavity Walls

A cavity wall consists of masonry built in two leaves or skins, with a void or cavity between them. Cavity walling began to replace solid walls for the external walls of domestic buildings during the latter parts of the nineteenth century but did not become common until the 1920s. The primary function of the cavity was to act as a barrier to prevent the passage of moisture from the outside to the interior of the building, the fundamental principle being that the inside and outside of external walls are kept separate, resulting in a dry interior. Initially the cavity tended to be very narrow but by the 1970s, the width had increased (to 50–75mm), as designers began to recognize the additional benefits of insulating the cavity to achieve a better balance of temperature for the inside environment of the building, keeping the interior cool in the summer and retaining heat in the winter.

Cavity insulation became compulsory under the Building Regulations in the 1990s and today's modern insulation requirements necessitate a cavity width of up to 100mm. Insulation materials are introduced into cavities during construction. The cavity is either partially or fully filled in order to improve the thermal performance of the wall and to meet regulatory requirements.

A typical cavity-wall cross-section shows the minimum dimensions in millimetres for cavity-wall construction. The thermal insulation requirements of the Building Regulations and practicality dictate that the external skin is typically constructed in facing brickwork and the inner skin in concrete blockwork that is at least 100mm thick.

The masonry below ground level can be constructed either of concrete blocks or of common bricks that have been designated as suitable for use below ground. The brickwork between finished ground level and the level of the horizontal DPC will be in contact with moisture from the ground, and this will make it vulnerable to frost damage in winter. In addition, any soluble salts present in the ground will be absorbed during periods of wet weather, resulting in white salt deposits forming on the face of the bricks (efflorescence) during dry weather. Externally, between finished ground level and horizontal DPC, therefore, it is

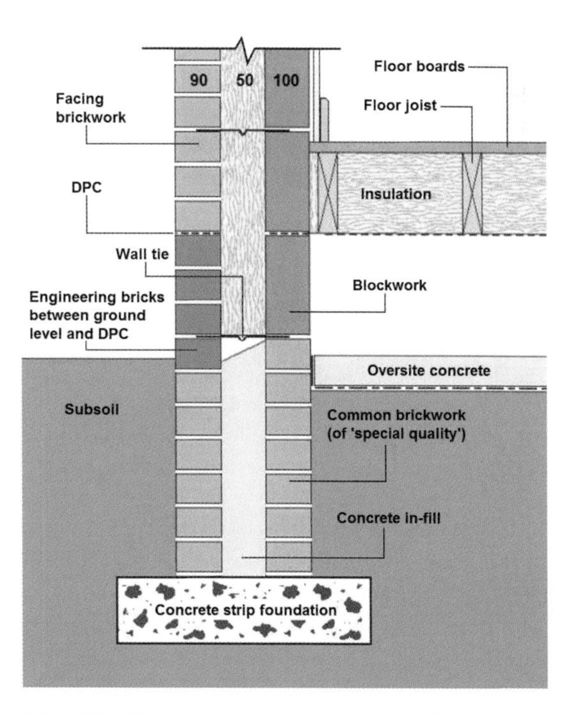

Fig. 167 Typical cavity-wall cross-section, showing the minimum requirement for its construction, in millimetres.

good practice to use a brick that is of engineering quality, with low levels of water absorbency. Facing brickwork starts immediately above DPC level.

Cavity walling below ground level must be capable of withstanding the lateral pressures exerted by the sub-soil. To this end, when the construction has reached ground level, the cavity is filled with a lean mix of concrete, which is then given a chamfered finish at the top in order to direct any moisture in the cavity towards the external leaf. Weepholes (see Figs 190 and 191) should be introduced into the external leaf at ground level at 890mm intervals (in other words, four bricks) to allow any moisture build-up to be expelled.

HORIZONTAL DAMP PROOF COURSE (DPC)

Capillary action can cause moisture to rise vertically through a structure to a height of around 1000mm. Cavity walls must have a horizontal barrier inserted at the base of the wall, to prevent moisture from the ground rising to a point where the internal living space becomes damp. A horizontal damp proof course (DPC) is positioned in both internal and external leaves of a cavity wall at a level that is no lower than 150mm above finished ground level, as required by the Building Regulations. The Building Regulations further require that the DPC be positioned flush with the face of the wall, within a tolerance of + or – 2mm. The damp proof course should also be free from tears and punctures and be uniformly bedded.

The most common material used to form a horizontal damp proof course is a flexible DPC felt, which is actually not a felt-like material at all but a thin, 'polymeric' sheet plastic supplied in rolls of varying widths from 100mm up to 600mm. Both sides of the felt have a fine, raised diamond pattern in order to provide a key, as the material is designed to be sandwiched within a 10mm bed joint of mortar. This is achieved by spreading a thin bed joint of mortar, rolling out the felt on top and then smoothing it out and flattening it using the bedding face of a brick. Care must be taken not to damage the felt during this part of the process. Another thin bed joint of mortar is spread on top of the felt for bedding the next course of bricks. Under no circumstances should the felt be laid directly on top of a course of brickwork with no mortar beneath it.

Fig. 168 DPC felt in different roll widths.

At corners and joints in straight runs, DPC felt must be overlapped by a minimum of either 100mm or the width of the felt, whichever is the greater. This minimum overlap is derived from the notion that water can travel up to 100mm horizontally by capillary attraction. It is good practice to exceed the minimum overlapping requirement wherever possible.

Staffordshire Blue Engineering Bricks

Due to their impervious nature, solid Staffordshire Blue engineering bricks laid in two courses around the base of a building provide an effective damp proof course on their own. However, modern Building Regulations requirements demand that a length of DPC felt must be laid over the top of the 'blues' as well, which rather negates their purpose.

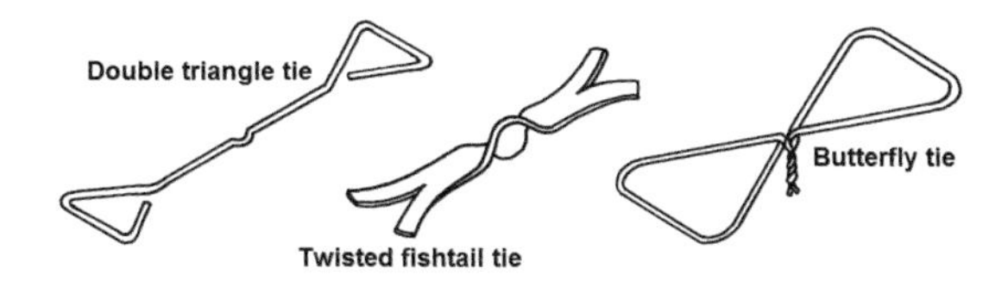

Fig. 169 Examples of traditional cavity-wall ties.

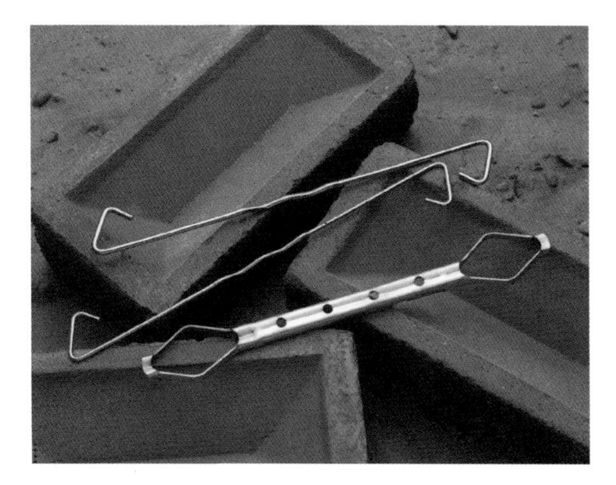

Fig. 170 Modern performance-type cavity-wall ties. Ancon Building Products

CAVITY-WALL TIES

Material and Shape

Wall ties bond the two leaves of a cavity wall together to make the two act as one structural wall. Those in general use are made of galvanized mild steel or of stainless steel, but more modern 'performance-type' ties of superior strength and quality are also available, and these have started to supersede the more traditional ties.

Ties come in various sizes to suit different cavity widths and are specially shaped so as to form a drip at their centre. If moisture penetrates the outer wall and passes along the wall tie it will meet this drip, fall into the cavity and seep away through weepholes placed at the base of the outer skin of the cavity wall, below the horizontal damp proof course. Weepholes are also found above door and window openings and any other areas where a cavity is bridged (for example, by an airbrick and associated ducting).

Positioning Wall Ties

When wall ties are bedded, they should not slope towards the inner leaf; instead they should be horizontal or, at worst, slope slightly towards the external leaf of the cavity wall. Wall ties must have a firm bedding of at least 50mm (62.5mm is recommended) into each leaf of the cavity wall, and the bedding should be even on both sides. When placing ties, they should be pushed into mortar beds after spreading to ensure that they are firmly fixed within the joints.

Wall ties are positioned 900mm horizontally, 450mm vertically (six courses of brickwork or two of blockwork) and 'staggered' over the face of the wall. Staggering ensures a consistency in the support provided by the ties, and allows for an even distribution of the load between the two leaves of the cavity wall. Without this approach, cavity walls would have vertical columns of great strength interspersed with structurally weak areas where there are no ties.

When openings for doors or windows occur they represent an effective point of weakness in the structure. Wall ties must be placed within 150mm horizontally of the reveal and every 225mm (three courses of brickwork or one of blockwork) in height near the sides of the openings. This is often

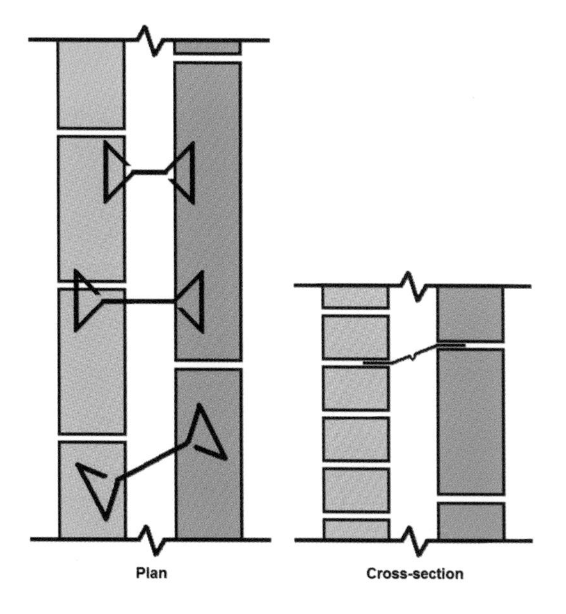

Fig. 171 Bedded wall ties.

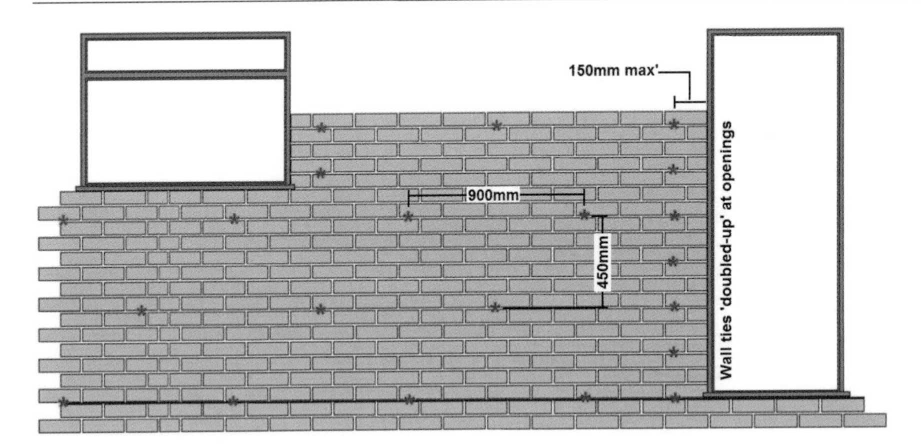

Fig. 172 Horizontal and vertical spacing of cavity-wall ties.

referred to as 'doubling up at openings' (see Fig 172).

CAVITY-WALL INSULATION

Cavity-wall insulation falls into two categories, based upon whether the insulation partially fills the width of the cavity (partial-fill) or completely fills the full width of the cavity (full-fill). The choice of insulation method depends on the nature of the materials that are selected to construct and insulate the cavity wall in order to meet the thermal insulation requirements of the current Building Regulations. From a bricklayer's point of view, the choice of insulation method is an important consideration as it determines the manner in which the cavity wall is constructed.

Full-Fill Insulation

Insulation designed to fill the full width of the cavity comes in 'batts' of lightweight, flexible insulation material (such as Rockwool), which are usually 450mm high by 1500mm long. Different thicknesses are available to suit different widths of cavities. The height is equivalent to six courses of brickwork (in other words, the vertical spacing of wall ties), so it

Fig. 173 Full-fill cavity-wall insulation.

Fig. 174 Installation of full-fill cavity insulation.

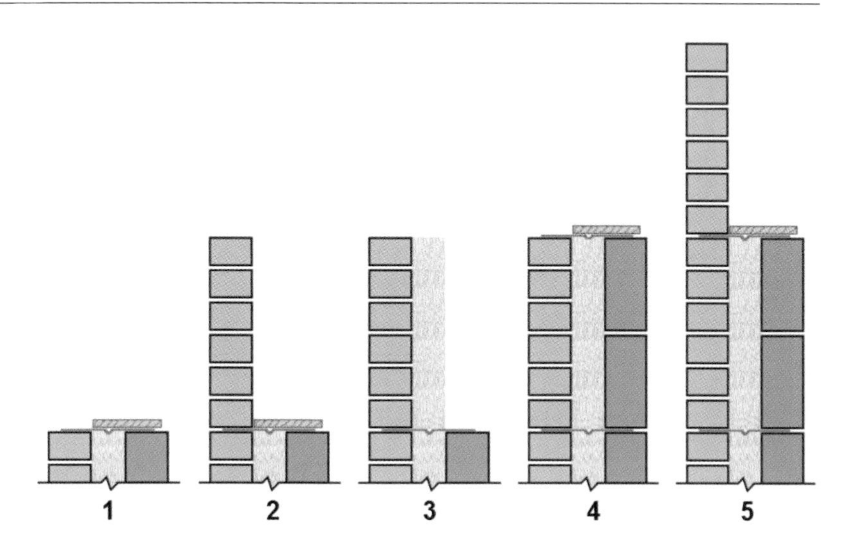

is usual to construct the external skin of brickwork six courses high, position the insulation at the back of the brickwork, then construct two courses of blockwork for the inner skin, before repeating the process again.

The notion of full-fill insulation, on the face of it, seems to defeat the whole object of introducing a cavity into an external wall in the first place, with the cavity itself being designed to act as a moisture barrier. Insulation designed for this application is manufactured with the aim of not allowing moisture to pass through, but this can never be completely guaranteed in locations exposed to severe weather conditions.

During the construction of each successive six courses of brickwork it is important to prevent mortar droppings collecting on top of the insulation slabs. If this waste is allowed to build up between the insulation slabs, it can form a 'bridge' for moisture to pass across the cavity to the inside of the finished building. The problem can be prevented by placing a timber board on top of the last course of blockwork to catch any mortar droppings that fall during construction of the next six courses of brickwork.

Partial-Fill Insulation

Rigid foam insulation, designed to partially fill the width of the cavity, provides the required thermal insulation properties but provides an effective barrier to rain penetration by maintaining the traditional cavity between the leaves. Insulation boards of 450mm × 1500mm must be fixed directly to the cavity side of the blockwork by way of proprietary plastic discs that clip on to the cavity-wall ties and retain the insulation in position (*see* Fig 176). This means that the blockwork needs to be constructed ahead of the brickwork, which follows on behind after the insulation has been clipped in place. Boards are available in thicknesses of 30mm up to 70mm, to suit cavity walls of different widths and to comply with the Building Regulations requirements for different residual cavity widths.

Fig. 175 Rigid foam insulation, or partial-fill cavity insulation.

Fig. 176 Plastic insulation-retaining clip.

Cavity Battens

If waste mortar is allowed to drop down the cavity and collect at the bottom, there is a risk of the horizontal DPC becoming bridged by the build-up. Similarly, mortar droppings should not be allowed to collect on the wall ties, as this defeats the object of the integral drip and could eventually lead to damp patches on interior wall surfaces. Obviously this is not an issue where full-fill cavity insulation is concerned but partial-fill insulation leaves a residual open cavity and great care must be taken to ensure that mortar does not fall down it during the construction process. The bricklayer needs to be very careful when spreading mortar bed joints and should trim off excess mortar when laying brick and blocks.

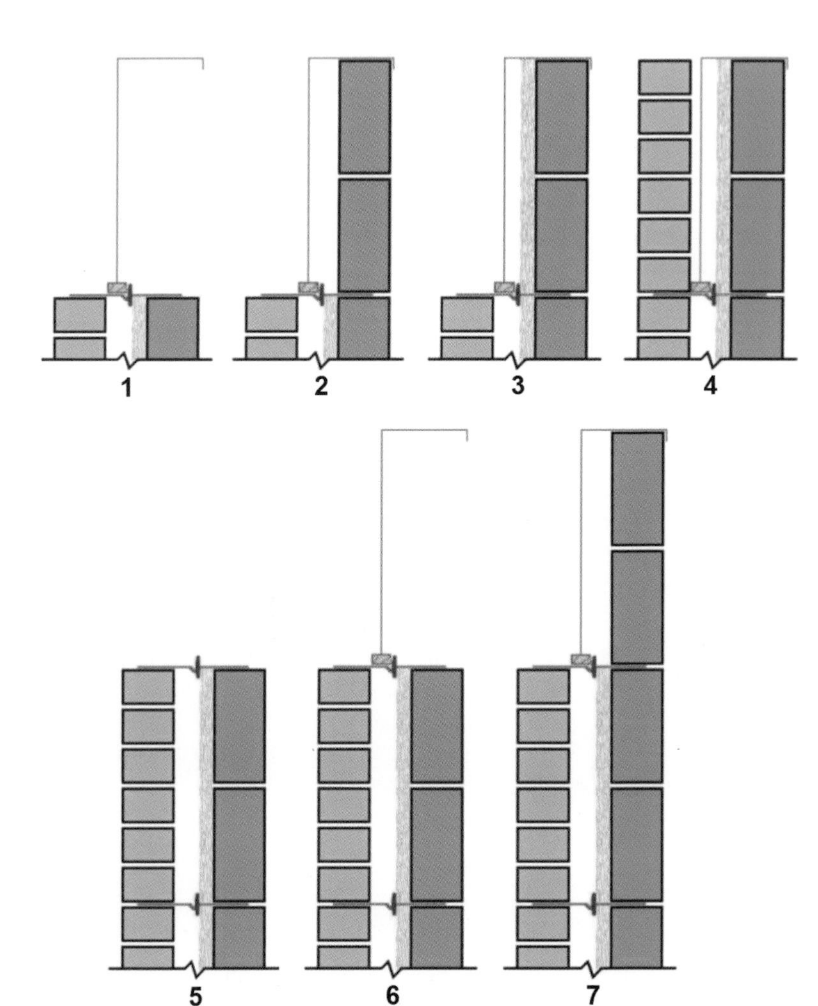

Fig. 177 Installation of partial-fill cavity insulation.

Being careful does not solve the problem completely, but mortar can also be prevented from dropping into the cavity by placing a timber batten on to the last horizontal row of wall ties (*see* Fig 177). The batten is slightly narrower than the residual cavity and wire is attached to it, which allows it to be raised as the next row of wall ties is placed. Each time, it is cleaned off before being repositioned on to the new row of wall ties. Where the residual cavity is narrow, it may be too difficult to use a cavity batten and the bricklayer will have to make a judgement as to whether it is worth the bother, given that there is a reduced chance of mortar falling down into the narrower cavity. Good practice, however, dictates that using a cavity batten is the preferred course of action.

ABOVE: **Fig. 178 Simple sill detail – uninsulated, resulting in 'cold bridging'.**

WINDOW SILLS AND DOOR THRESHOLDS

Sealing the Cavity

The sill is the lower horizontal portion of a window opening and the threshold is the corresponding part of a door opening. The cavity at sill and threshold level is sealed by turning the internal blockwork across the cavity into the back of the external leaf of brickwork. A length of DPC felt (typically 200mm wide) is inserted at the back of the brickwork and turned over the top of the brickwork under the sill before positioning the frame. The length of felt should be sufficient so that it extends beyond the width of the opening by 150mm at both ends. This will ensure an adequate 100mm overlap with any vertical DPC felt placed in both reveals of the opening. The length of felt

BELOW: **Fig. 179 Simple sill detail – insulated, to prevent 'cold bridging'.**

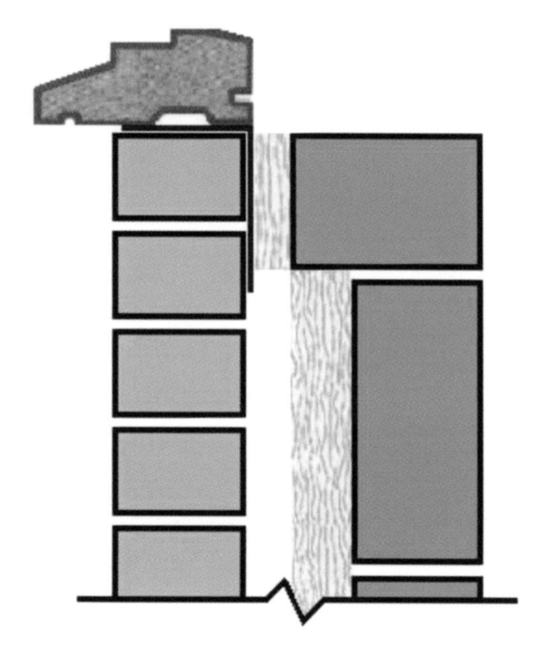

Damp Proofing at Openings

It is vitally important around openings where the cavity is sealed with blockwork (and the cavity bridged) that an effective and continuous means of damp proofing is introduced into the construction, to maintain the integrity of the cavity and its purpose for being there. This applies as much to reveals and heads of openings as it does to sills and thresholds.

used is, therefore, equal to the width of the opening plus 300mm.

Unless aerated concrete blocks of the correct thermal grade are used, this form of construction is not regarded as good practice today because of the risk of 'cold bridging', which could cause internal condensation. To prevent cold bridging, insulation board or insulated DPC is introduced at the point where the blockwork seals across the cavity at sill level.

The most common and simplest sill detail is an integral sill where the sill forms the bottom part of the window frame. The groove at the front of the timber sill is called a 'throating' and acts as a drip to help prevent water running back under the sill and/or down the face of the brickwork. The drip encourages a build-up of water, the weight of which eventually overcomes its own surface tension, causing it to drop to the ground. The throating must, therefore, be positioned well clear of the wall. The throating of a sill works in much the same way as the drip on a cavity-wall tie. The sill of the frame is bedded in mortar on to the external leaf of brickwork in order to ensure level and even bearing on the brickwork.

Sub-Sills

An alternative to sills that are part of the frame inserted into an opening is to construct a structural sub-sill that provides a weathering at the bottom of the opening. A typical example is the use of a brick-on-edge sub-sill, which projects beyond the face of the main wall and is sloped so as to allow rainwater to fall clear of the wall below. Sometimes, specially shaped bricks such as 'bull-nose' or 'cant' bricks are used to make the sill detail more decorative. Adequate slope and projection is needed as it is not possible to create a throating under a brick-on-edge sub-sill.

A DPC must be placed under the brick-on-edge, which is turned up at the point where the cavity is sealed to form a tray. The length of felt should be sufficient to extend beyond the width of the opening by 250mm at both ends. This will ensure an adequate 100mm overlap with any vertical DPC felt placed in both reveals of the opening, as well

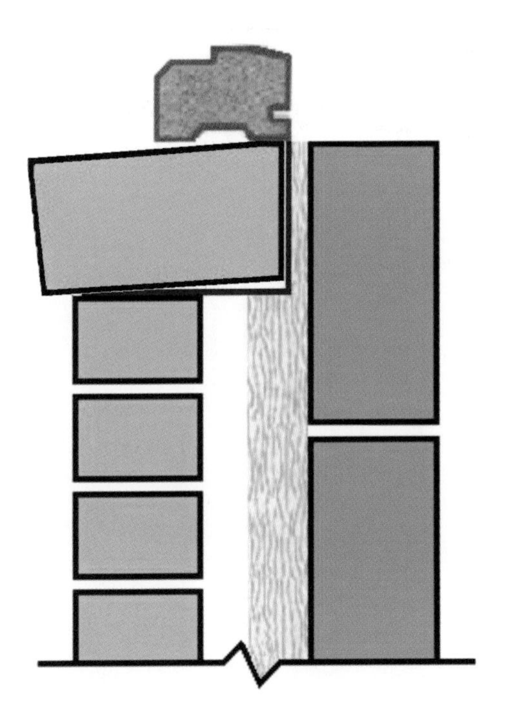

Fig. 180 Sill detail incorporating a brick-on-edge sub-sill.

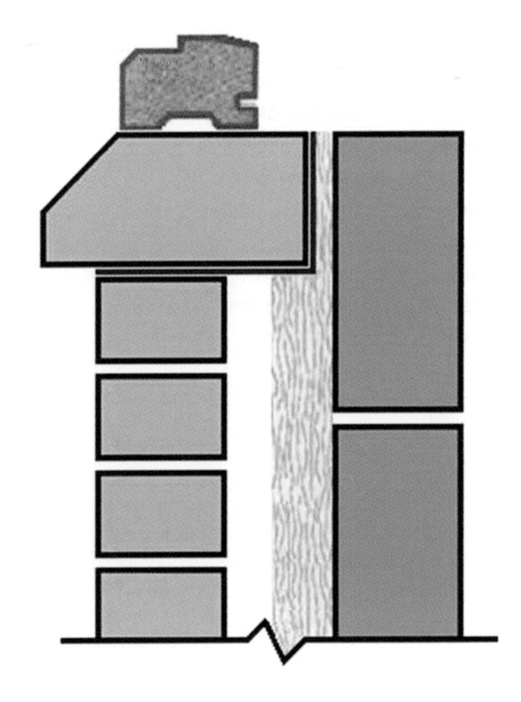

Fig. 181 Sill detail incorporating a 'cant' brick sub-sill.

as projecting by a further 150mm into the cavity at both ends. The length of felt used is, therefore, equal to the width of the opening plus 500mm. Alternatively, the tray can be provided with proprietary stop-ends or cloaks, which negate the need to extend the tray beyond the reveal and into the cavity.

In areas of severe or very severe exposure to rain, and especially when full-fill cavity insulation has been employed, it is a requirement to provide proprietary cloaks or stop-ends to cavity trays at sills instead of letting the tray terminate open-ended where it runs past the ends of the sill, into the vertical reveal and beyond into the cavity.

Insulation board is inserted where the blockwork closes off the cavity to prevent cold bridging. Alternatively, an insulated plastic cavity closer could be used (see Fig 184).

DOOR AND WINDOW REVEALS

The term 'reveal' refers to the vertical inside face of openings against which door and window frames are fixed. Cavities are closed at reveals by returning the internal blockwork across the cavity into the back of the external leaf of brickwork (see Fig 182). A vertical DPC felt (150mm wide) is inserted at the back of the brickwork, projecting into the cavity by 25mm and out of the reveal by 25mm. It is folded forward and 'trapped' between the reveal and the door or window frame. The length of felt must be sufficient that it projects 150mm below sill level and overlaps with any DPC felt or tray positioned in the sill construction. In addition, the felt must project above the reveal by at least 25mm so it can be folded forward and 'trapped' under the bearing end of the lintel used to bridge over the door/window opening.

Again, this form of construction is not regarded as good practice today because of the risk of 'cold bridging', unless aerated concrete blocks of the correct thermal grade are used. To prevent cold bridging, insulation board or insulated DPC is introduced at the point where the blockwork seals across the cavity at the reveal.

A more recent development is the insulated plastic cavity closer, which comes in a variety of widths to suit different sizes of cavity and is easily cut to length on site (see Fig 184). Cavities are left

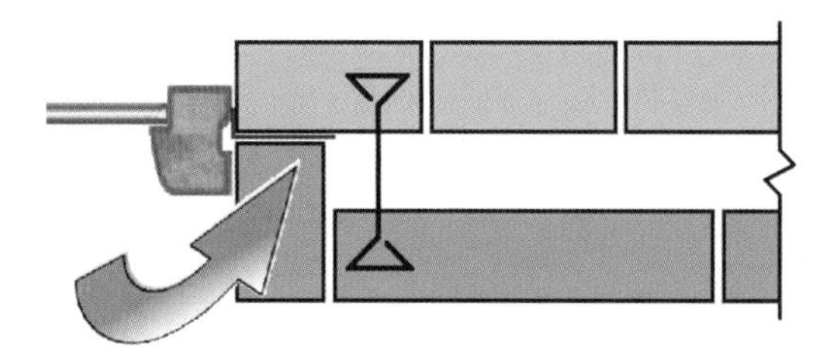

Fig. 182 Simple reveal detail – uninsulated, resulting in 'cold bridging'.

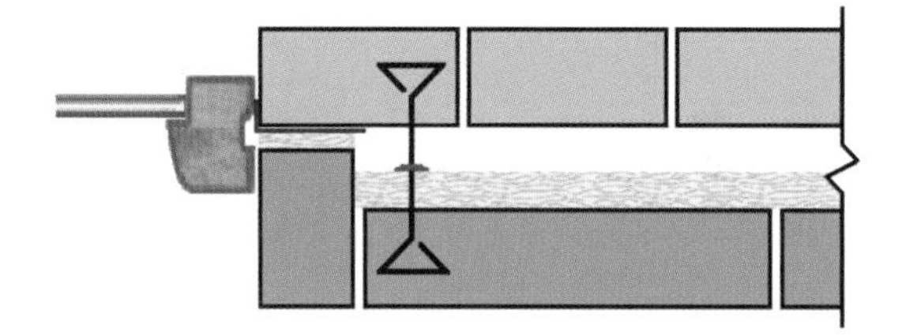

Fig. 183 Simple reveal detail – insulated.

Fig. 184 Insulated cavity closer installed at a window reveal.

open at sills and reveals to allow the cavity closer to be inserted. Not only does the closer prevent cold bridging and seal the cavity, its extruded plastic construction acts as a DPC. This is a much more efficient and speedy form of construction, which eliminates complex masonry detailing and the insertion of DPC felt.

BRIDGING OPENINGS IN CAVITY WALLS

To allow construction to carry on over the top of an opening in a wall, the opening must be bridged with a lintel that is wider than the opening, allowing the ends to bear on to the walling either side of the opening. The purpose of a lintel is to support the weight of the masonry above the opening and to distribute that load through the height and width of the lintel down into the brickwork either side, upon which the lintel ends bear.

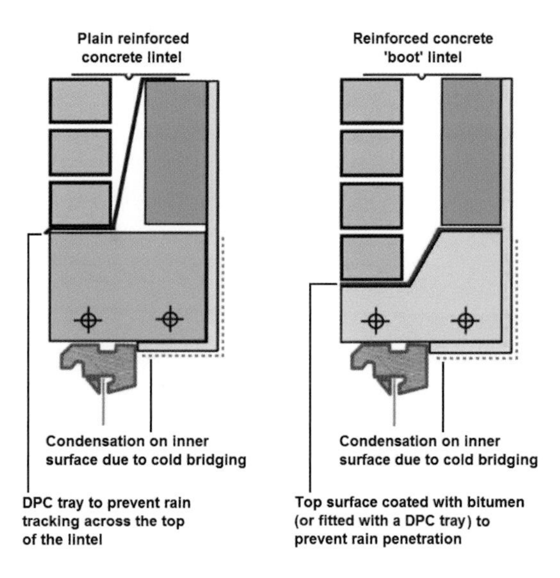

Fig. 185 Simple pre-cast reinforced concrete lintels.

Concrete Lintels

In older domestic construction, openings were bridged by using reinforced concrete, in the form of either a pre-cast lintel or one that is cast in-situ. For the latter, timber or metal formwork is erected around the top, or head, of the opening and wet concrete is poured into it to form the lintel. In both cases, the lintel will have steel reinforcing bars in the bottom portion to overcome the weakness of concrete under tension. Where concrete lintels are used today they are invariably pre-cast, due to the length of time taken to cast concrete in-situ and the consequent delay to the construction process.

Concrete lintels tend not to be used in the external walls of new domestic buildings, as they have a number of disadvantages:

- compared to steel, they are costly and are heavy to lift into position (often requiring lifting machinery);
- they have poor thermal insulating qualities, resulting in condensation on the inside around the top of the window opening;
- they allow rainwater to track across to the inside;
- they are available only in a limited range of sizes;
- they cannot span very wide openings; and
- they are not particularly attractive to look at.

To prevent moisture inside the cavity from tracking across to the internal leaf, a DPC tray is introduced over the top of the lintel, to deflect any water in the cavity towards the external leaf. The tray is formed using a length of felt, with the front edge positioned in the bed joint directly above the lintel on the outer leaf of the cavity wall. It is turned up the cavity to form a slope and its back edge is bedded in a higher mortar joint in the blockwork of the inner leaf of the cavity wall. The length of the felt should be sufficient so that the cavity tray projects into the cavity, beyond each end of the lintel, by at least 150mm. Accordingly, the felt needs to be as long as the lintel, plus 300mm. Alternatively, the tray can be provided with proprietary stop-ends or cloaks, in which case it will not need to be extended beyond the ends of the lintel.

In areas of severe or very severe exposure to rain, and especially when full-fill cavity insulation has been employed, it is a requirement to provide proprietary cloaks or stop-ends to cavity trays above lintels. The tray must not be allowed to terminate open-ended where it runs past the ends of the lintel into the cavity.

The width of felt used to form a cavity tray will depend somewhat on the width of cavity but, to ensure that the felt is adequately bedded into each leaf of the cavity wall, a width of 450mm is usually sufficient. Weepholes must always be provided in the external leaf of cavity walls, in the course of brickwork beneath which the DPC tray is bedded, in order to provide an escape route for any water in the cavity.

Concrete lintels are difficult to disguise or make visually pleasing within facing brickwork, so they tend to be limited to bridging openings in blockwork internal partition walls where they are not seen. One way of attempting to improve the aesthetics was to use a pre-cast 'boot' lintel (so called because of its shape in cross-section), which had a reduced depth at the front so that less concrete was visible. The shape and design of more modern steel lintels reference the concrete boot lintel. Pre-cast concrete lintels require an end-bearing on the masonry either side of the opening of at least 100mm on a bed of mortar.

Steel Lintels

Steel lintels (*see* Fig 186) for bridging openings in external cavity walls are commonly made of either stainless steel or powder-coated galvanized steel. They come in numerous sizes, sectional shapes and lengths to suit various thickness of cavity wall and width of opening. Heavy-duty versions are capable of spanning clear openings of over 6m. Steel lintels typically have an end-bearing on the masonry either side of the opening of 150mm on a very thin bed of mortar in order to allow brickwork to be continued across the front of the lintel. The advantages of steel over concrete are that it is cheaper, lighter, and easier to position, allowing for faster construction. A steel lintel is also more visually appealing than a concrete one, as the end-bearing is concealed within a mortar bed joint and only a narrow lip of steel is visible externally over the top of an opening.

The sloping profile or cross-sectional shape of the lintel is a deliberate design feature that means the lintel forms its own DPC tray. A proprietary stop-end or cloak is inserted at each end, which forms a vertical up-stand and provides an additional measure to ensure that any water in the cavity is directed toward the external leaf. A separate cavity tray of felt is not normally required but some consider it good practice to include one. Again, the piece of felt should be long enough so that the cavity tray projects into the cavity at least 150mm beyond each end of the lintel. The length of felt required is therefore equal to the length of the lintel plus 300mm. Alternatively, the tray can be provided with proprietary stop-ends or cloaks and it will not then need to be extended beyond the ends of the lintel.

Again, in areas of severe or very severe exposure to rain, and especially when full-fill cavity insulation has been employed, proprietary cloaks or stop-ends must be provided to cavity trays above lintels, instead of letting the tray terminate open-ended where it runs past the ends of the lintel into the cavity.

To avoid a 'cold bridge' through the cavity wall, steel lintels must incorporate continuous insulation along the length of any box section. Without insulation, 'cold spots' would form on the inside wall above the top of the opening, and on the underside or soffit of the opening beneath the lintel, which could lead to condensation forming.

Lintel Reinforcement

When a concrete beam or lintel is under load it will sag. This causes the top half of the lintel to shorten and to be 'in compression', with the bottom half essentially lengthening as it is subjected to tensile stretching forces. Concrete is incredibly strong under compression but very weak under tension, so concrete lintels require steel reinforcing bars to be installed in the lower section during the casting process. The reinforcement has a ribbed surface and 'hooked' ends, so as to remain solidly gripped within the concrete that surrounds it. When steel reinforcement is cast inside concrete, it must have a minimum of 50mm of concrete around it. This concrete cover protects the steel from moisture in the atmosphere and penetrating rain. The steel must not be allowed to rust, since steel expands when it rusts and will crack or otherwise damage the concrete, causing the component to fail. Accordingly, a lintel must be installed the right way up in order to avoid structural failure. Manufacturers usually mark the top of a lintel for this purpose.

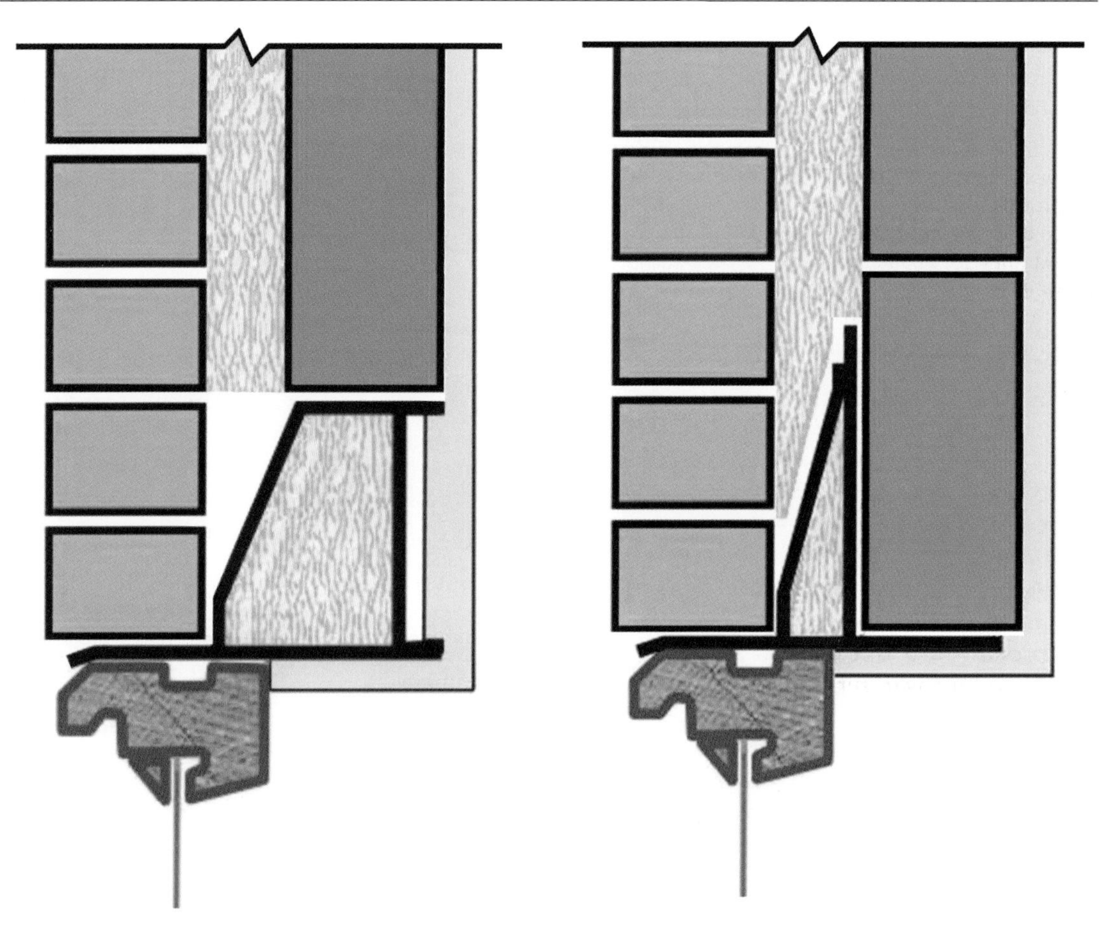

Fig. 186 Insulated box lintel.

Fig. 187 Insulated open-backed lintel.

An alternative to the box-section steel lintel is the open-backed lintel, which has a number of advantages. Blockwork, as opposed to steel at the back, provides a better key for plaster. In addition, there is no requirement for blockwork to be cut round a box section and such lintels can be utilized where fair-faced blockwork (or brickwork for that matter) is desired internally.

AIRBRICKS

Airbricks provide perforated openings built into the base and all round the external walls of a domestic building, in order to ventilate the space beneath a suspended timber floor. They are typically made of ceramic, metal or plastic. Without them, there is

Fig. 188 A selection of airbricks.

Cavity Trays over Airbricks

In the event that airbricks are built into a cavity wall above DPC level, it is good practice to provide a cavity tray over the top of the airbrick so that any water in the cavity is diverted to the external leaf and not the inner leaf of the cavity wall. The installation follows exactly the same principles that apply to lintels above door and window openings.

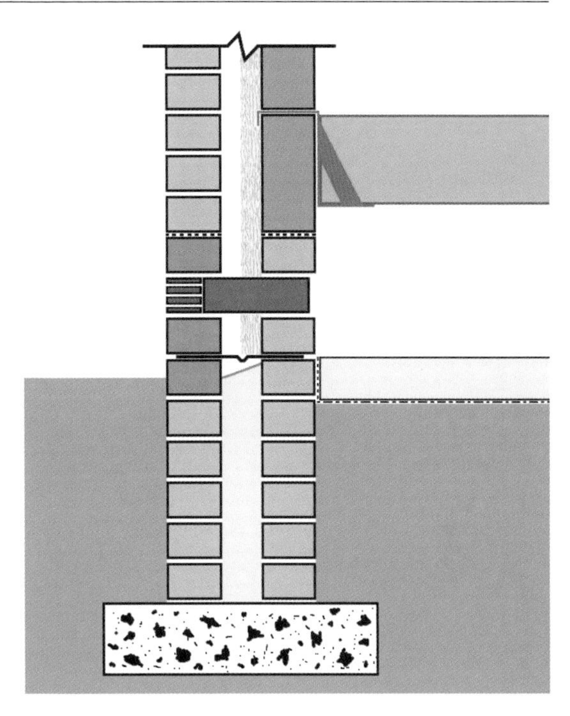

Fig. 189 Airbrick detail.

a strong possibility that lack of ventilation could result in the moisture content of the timber floor reaching a level at which 'dry rot' could commence. Airbricks are 215mm wide and are available in heights of 65mm, 140mm and 215mm (in other words, one, two or three courses), and in different colours to tone with surrounding brickwork. Plastic airbricks are purposely designed so they can be interlocked together to form larger units. Horizontal spacing of airbricks around the base of a building is determined by the Building Regulations.

Airbricks built into cavity walls must bridge across the cavity and direct the air flow straight under the floor. This is achieved with either a terracotta or plastic telescopic liner of matching width and height being positioned directly behind the airbrick. A plastic telescopic liner has the advantage that it can be adjusted to suit different widths of cavity wall. Liners channel the air straight into the under-floor space rather than displacing it into the cavity.

WEEPHOLES

There is always a risk that water can penetrate the external leaf of a cavity wall and find its way into the cavity. This risk is greater in geographical areas

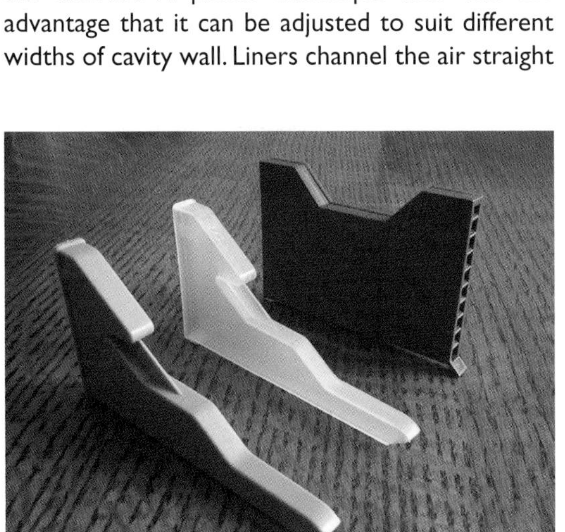

Fig. 190 Examples of plastic weephole inserts.

Fig. 191 Plastic weepholes inserted above a door opening.

Requirements for Weepholes

The requirement to provide weepholes is a comparatively recent development and very few bricklayers, if any, have claimed to see water exuding from a weephole. Many, therefore, question both the amount of water that finds its way into the cavity and the need for installing weepholes at all!

that are exposed to harsh weather conditions and/or where a more porous brick has been used for the external leaf of the cavity wall. As the cavity exists as a moisture barrier, it is vital that water does not cross the cavity and penetrate the internal leaf of the cavity wall.

Wall ties are provided with a 'drip' to ensure that water falls off into the cavity and all bridging points across a cavity such as lintels and airbricks are provided with a DPC tray, which deflects water back towards the external leaf. Another bridging point is where the cavity is filled below ground with a lean mix of concrete and finished with a sloping surface (towards the outside), at finished ground level. Around the base of a building, weepholes should be installed and horizontally spaced every 890mm. Above cavity trays over the tops of door or window openings, they should be spaced every 450mm.

Traditionally, to form a weephole, an open cross-joint would be left, but this tended to look untidy, was sometimes ineffective if the open joint was not completely clear of mortar, and also often got forgotten anyway. Nowadays, proprietary plastic inserts designed to fit into a cross-joint provide a much more satisfactory solution. They are available in a range of colours, to tone with the bricks being used.

CAVITY WALLS AT EAVES LEVEL

The term 'eaves' refers to the lower edge of the roof and includes the soffit and fascia boards. The most structurally sound treatment for termination of a cavity wall at eaves level is to seal the cavity with blockwork laid flat. This gives the wall greater stability, by tying the two leaves together, and improved load distribution over both leaves of the cavity wall (see Fig 192). As there is a risk of cold bridging at this point, aerated concrete blocks of the appropriate thermal grade should be used.

The timber wall plate (typically 100mm × 75mm sectional size) is bedded on mortar in order to compensate for any deviation along its length. It is then held down and secured to the inside face of blockwork with 'L'-shaped galvanized steel straps.

Ventilation into the roof space is achieved by vents left in the soffit board and plastic ventilation trays fixed between the timber rafters.

A simpler alternative treatment for the termination of a cavity wall at eaves level, which also eliminates the risk of cold bridging, is to omit the sealing course of blocks. The cavity is, therefore, left open at the top, with a timber wall plate positioned on top of the last course of blockwork upon which the roof timbers are fixed. This allows the loft roll insulation to be effectively joined up with the cavity-wall insulation, making a complete thermal barrier. The downside is that the top of the wall is not as structurally stable as a result of omitting the blockwork sealing course.

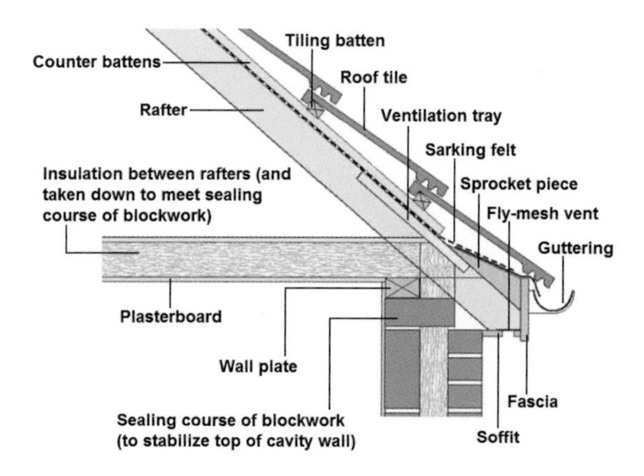

Fig. 192 Termination of a cavity wall at eaves level.

CHAPTER 12

Boundary Walls, Copings and Caps

BOUNDARY AND GARDEN WALLS

It is desirable for the sake of economy and time to build boundary and garden walls with relatively thin walling. The problem with such free-standing thin walls is their inherent lack of strength, so they need to be thickened with piers. These are attached at strategic points along the length of the wall, or where additional strength is needed, such as the point where a gate is to be supported.

Attached piers along the length of the wall provide increased strength against lateral forces (for example, wind pressure) that might otherwise overturn the wall. The inclusion of attached piers, their size and frequency along the wall length are directly affected by the height of the wall; the taller the wall, the greater the face area, and the more significant the lateral wind loads that have to be resisted. Tall, slender walls will be more vulnerable, as their height increases in relation to their thickness.

Legal Responsibilities

Building a wall to personal specifications carries with its certain legal responsibilities, particularly if the wall is adjacent to a public right of way or highway. Such responsibilities are often only considered when something goes wrong and a wall collapses, causing injury or worse. Quite simply, if there is any doubt as to the best way forward when designing a wall, particularly a complex one, the services of a structural engineer should always be engaged.

As a general rule of thumb, attached piers are spaced every 3m or so between end piers and/or returns (corners), which act as a buttress.

Walls higher than 2m start to enter the realms of structural design, including the provision of buttressing returns, and this is beyond the scope of this book. The advice here relates only to the more basic, lower structures.

ATTACHED PIERS

One-brick walls built in stretcher bond have an inherent weakness due to the continuous vertical 'collar joint' that runs the full height of the wall between the front half-brick thickness and the back half-thickness. Whilst this can be overcome by using short cavity-wall ties, stretcher bond is arguably best avoided for one-brick thick boundary walls. English and Flemish bonds are much stronger alternatives.

For attached piers to be effective, they must be bonded or tied into the main wall, be at least 100mm thick and, as a minimum, terminate within a height measured from the top of the wall equal to three times the wall's least thickness. For example, a pier attached to a one-brick wall (215mm thick) must terminate no lower than 645mm from the top of the wall.

More detail on finishing off the tops of boundary walls and piers is given later in this chapter.

Bonding Attached Piers

Since such piers are defined as being attached to the main wall, it follows that the main wall should be bonded first and then the projecting pier tied or bonded into it. This should be done in the course

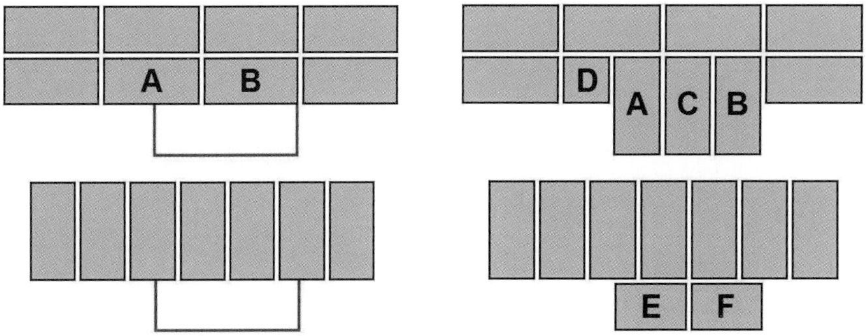

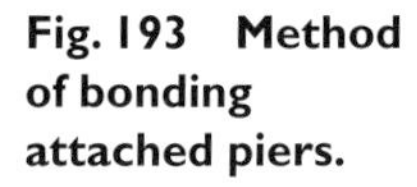

Fig. 193 Method of bonding attached piers.

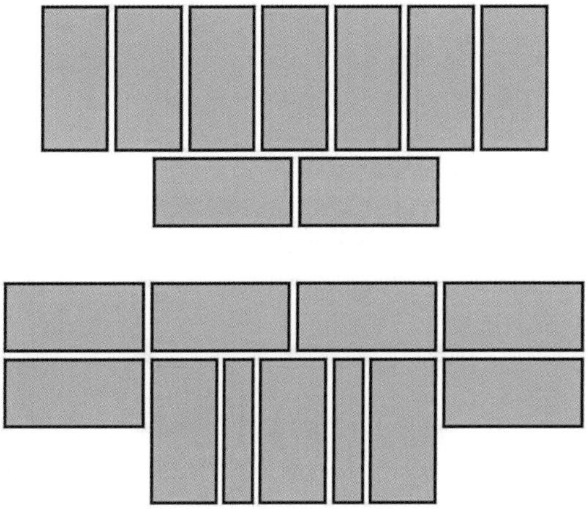

Fig. 194 English bond, two-brick wide single attached pier.

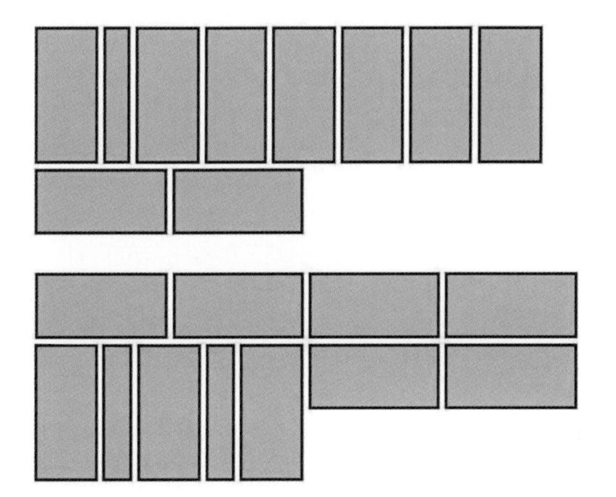

Fig. 195 English bond, two-brick wide single end pier.

that is most convenient while at the same time trying to adhere to the basic principles of bonding. When working out how best to tie in a pier to the main wall, particular attention should be given to the arrangement of bricks to ensure continuity of the transverse joints through both wall and pier, and to reduce cutting to a minimum.

Single Attached Piers in English bond

Single attached piers project on one face of the wall only. The example in Fig 193 shows the outline of a 1½-brick wide pier to be attached to an English bond wall (left), without any consideration given to 'bonding in' the pier. By removing bricks 'A' and 'B' and repositioning them (right), a bond for the pier into the main wall is achieved without significantly affecting the bonding of the main wall. The bond is completed by introducing header 'C' and half-brick 'D'.

It is structurally necessary to tie the pier into the main wall only on alternate courses (in other words, on the stretcher course), so all that remains is to complete alternate courses of the pier by arranging bricks to maintain a suitable face bond on the front of the pier. In the example, this is achieved by two three-quarter bricks, 'E' and 'F'. Figs 194 and 195 illustrate the bonding arrangements for two-brick wide single attached piers in English bond.

Single Attached Piers in Flemish Bond

The principles of bonding attached piers in Flemish bond is approached in much the same way as for English bond but, for the sake of convenience and to avoid excessive cutting, the face bond of

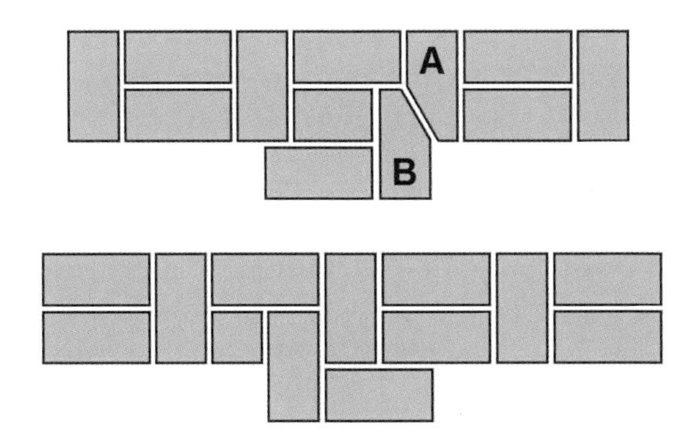

Fig. 196 Flemish bond, half-brick wide single attached pier.

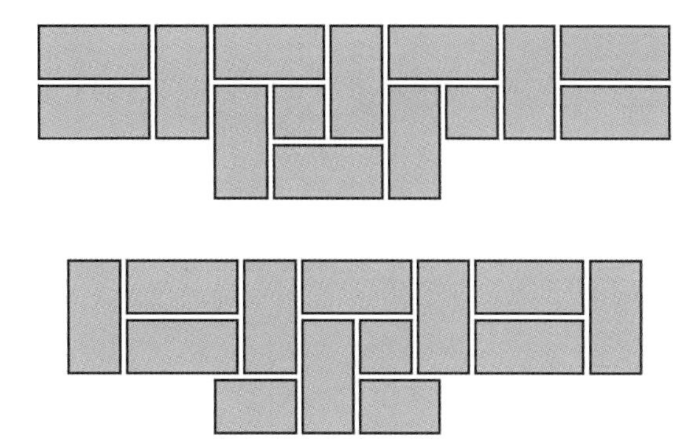

Fig. 197 Flemish bond, two-brick wide single attached pier.

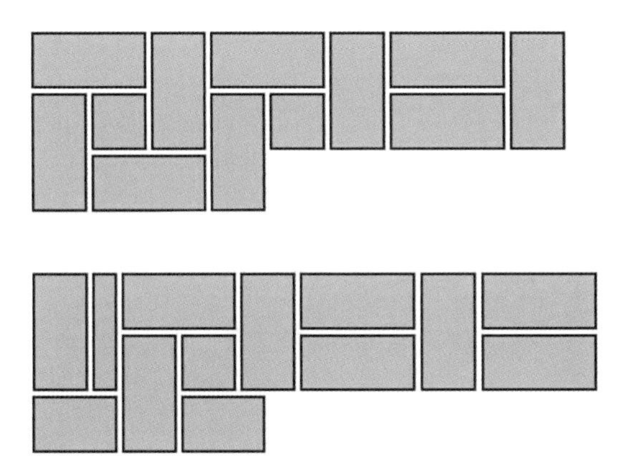

Fig. 198 Flemish bond, two-brick wide single end pier.

the pier usually ends up as English bond. Flemish bond for the main wall, being alternating headers and stretchers in the same course, also means that piers attached to Flemish bond walls are tied in wherever stretchers occur in the main wall so the pier becomes tied in on every course and not just alternate courses.

On the example shown in Fig 196, spayed cuts ('A' and 'B') known as 'King Closers' have been introduced on alternate courses. Placing a full header in the main wall (at position 'A') and merely placing a half-brick on the right-hand side of the pier (at position 'B') would correctly maintain the face bond, but would result in a continuous vertical joint throughout the height of the wall and the pier not being tied in on every course.

Double Attached Piers in English bond

Further increased stability can be achieved for free-standing walls by doubling the size of the attached pier, so that it projects on both sides of the wall (see Figs 199 and 200).

The principles of bonding are the same as for a single attached pier, where the main wall is bonded first and then a single attached pier is tied in to it. This is then simply repeated on the other side of

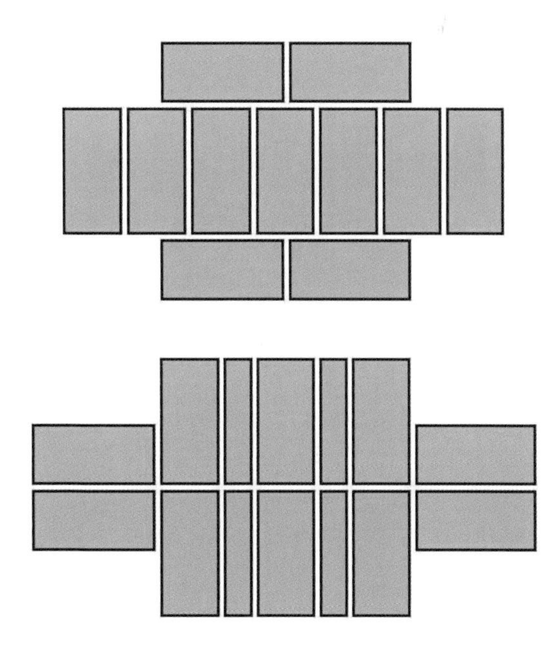

Fig. 199 English bond, two-brick wide double attached pier.

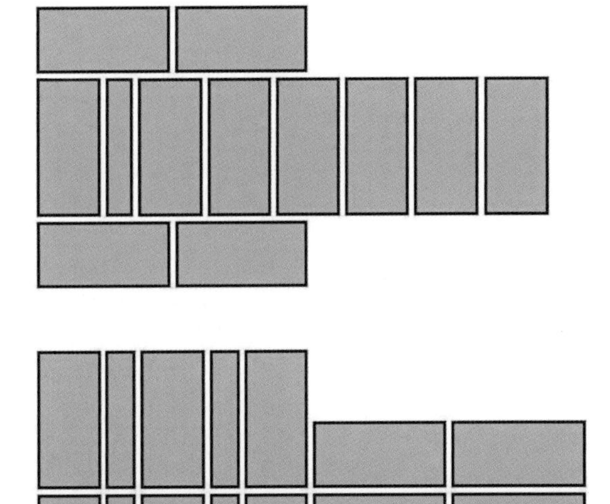

Fig. 200 English bond, two-brick wide double end pier.

the wall, to form a double pier. The double attached pier can be thought of as two single attached piers positioned on opposite sides of the wall, at the same relative point along the wall length.

Double Attached Piers in Flemish Bond

The bonding of the double pier in Flemish bond (see Fig 201) clearly requires a lot of brick cutting, but it is possible to simplify the bonding arrangement and reduce the amount of cutting involved. The alternative arrangement (see Fig 202) completely goes against the principles of bonding previously described; this example would require the pier bond to be set-out first. Simplifying the bond does increase the number of vertical straight joints within the pier, which reduces its strength,

> **Running-in Walls with Attached Piers**
>
> A boundary wall that is long enough to incorporate attached piers in the middle will be run-in to a string-line. When this is done, the attached pier is always kept one course behind the main wall – that is to say, each course of the wall is run-in to the line, then the line is removed so that the corresponding course of the attached pier can be completed. Constructing the wall in any other order would mean that the pier brickwork would interfere with the positioning of the string-line, preventing the main wall being run-in in one go.

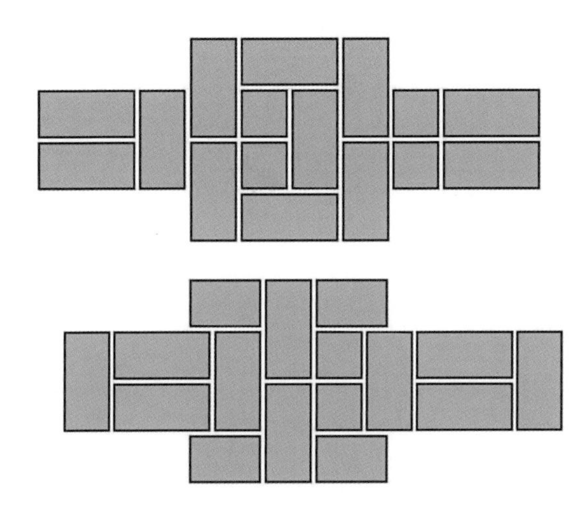

Fig. 201 Flemish bond, two-brick wide double attached pier.

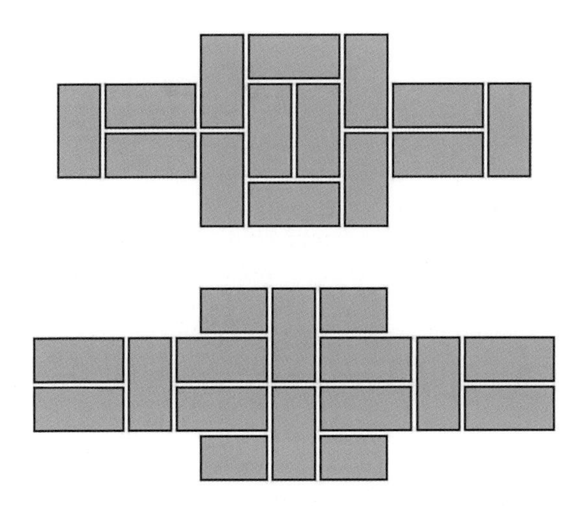

Fig. 202 Flemish bond, two-brick wide double attached pier (alternative, simpler bonding arrangement).

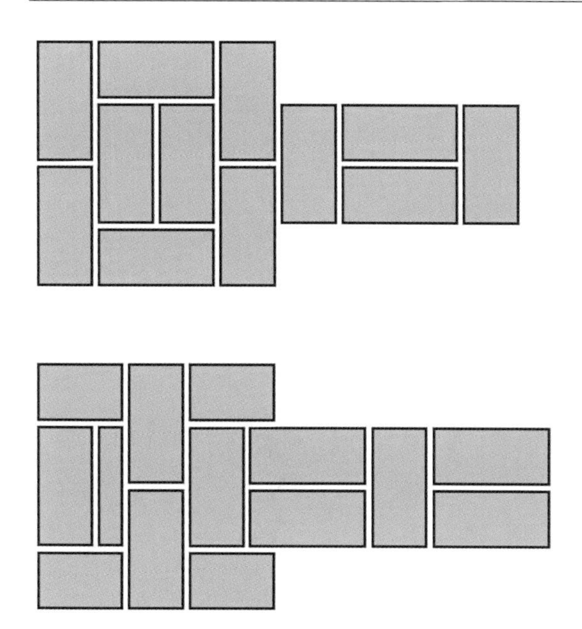

Fig. 203 **Flemish bond two-brick wide double end pier.**

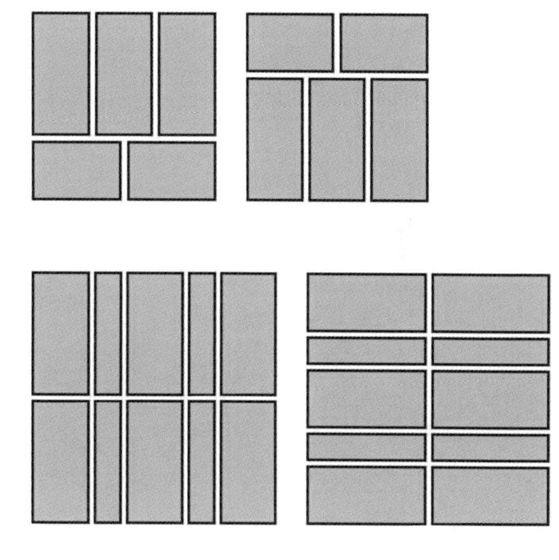

Fig. 204 **English bond isolated piers.**

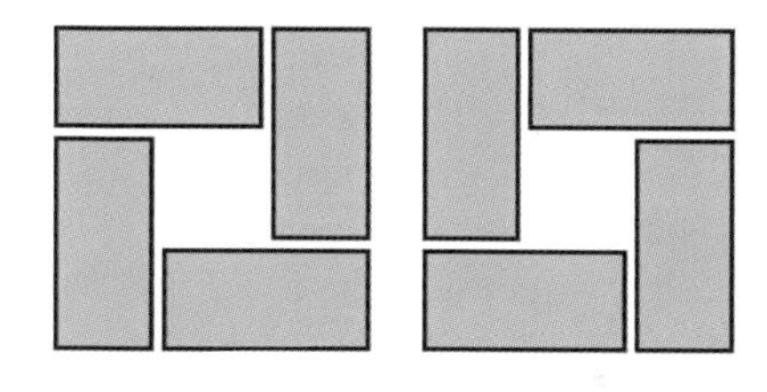

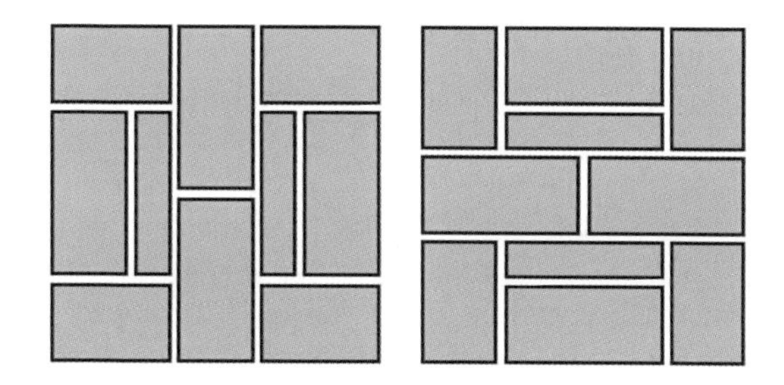

Fig. 205 **Flemish bond isolated piers.**

but not to the point where the purpose of the pier is significantly compromised or undermined. Strictly speaking, in terms of the principles of bonding, this alternative approach is incorrect, but the bond does 'work' and the pier still does its job. As a result, it does tend to be more common practice on site.

ISOLATED PIERS

An isolated pier or pillar is the opposite of an attached pier in that it stands on its own with no connection to, or support from, any other walling. An isolated pier may be constructed as a support for a beam, as a gate pillar or as a decorative feature to provide a base for an ornamental feature such as a sundial, bird bath or statue.

In order to be structurally stable a pier must not be too slender and, in practical terms, the height of an isolated pier must not exceed eight times its least thickness. Also, a pier ceases to be regarded as such when its effective length exceeds four times its effective thickness; after this, it becomes classified as a wall.

Constructing Isolated Piers

When building the pier, it is fundamentally important for it to be vertically upright ('plumb') and for all four sides to be square with one another. The key to achieving the latter begins with the first course, which will dictate everything that happens on subsequent courses. The example (*see* Fig 206) is a 2-brick by 1½-brick pier in stretcher bond; following the correct sequence of operations will ensure that it is constructed square and plumb:

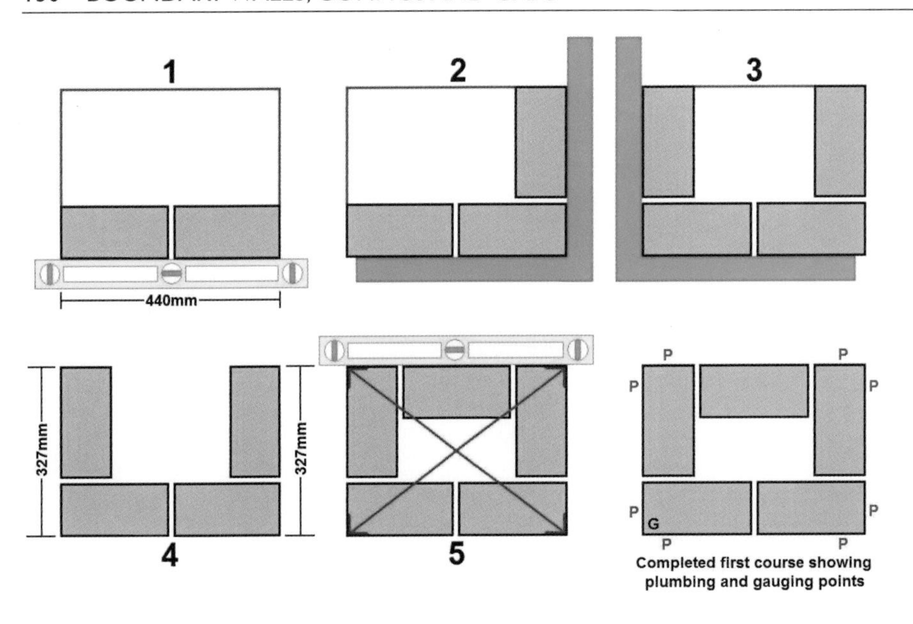

Fig. 206 Method of constructing an isolated pier.

1. Start by laying the front two stretchers to gauge, level, in line and to the correct length, in this case, 440mm (Stage 1).
2. Construct the two sides by squaring the side bricks from the front bricks, which are known to be in line (Stages 2 and 3).
3. Check that each of the two new sides measures 327mm (Stage 4).
4. The back brick can be simply 'run-in' to the ends of the side bricks by using the spirit level as a straight-edge.
5. As a final check, measure from corner to corner across both diagonals, which, if the first course is square, should be equal (Stage 5). A centre in-fill brick can be laid, if required, on each course.

Subsequent courses begin with the corner brick at the gauging point ('G') from which the remaining three corner bricks are levelled. When laid, each corner brick must be plumbed on both sides to ensure it is vertical and not twisting as it gains height, meaning that the pier has a total of eight plumbing points ('P'). After all the corner bricks have been positioned, the side bricks can be run-in for line and level to the corner bricks by using the spirit level as a straight-edge. Again, and on every course, check the diagonals to make sure the pier remains square.

When building piers of any type, attached or isolated, it is necessary to be selective about the bricks that are being used. The comparatively small size of a pier, the relatively small number of cross-joints and the need for great accuracy means that tolerances become quite tight. Any bricks that are oversized, undersized and/or misshapen should be discarded in favour of more consistently sized and shaped examples.

Sometimes, to save bricks, mortar and time, piers are constructed hollow. This is achieved by omitting the centre in-fill bricks and/or by reducing the length of bricks so that what remains is a hollow core surrounded by a half-brick skin. In some cases where the pier is not required for structural support, the void is left empty, but in other cases, where structural strength is required, the void is filled with wet concrete, sometimes supplemented with steel reinforcement. As wet concrete exerts a

Plumbing Isolated Piers

A tell-tale sign that a pier is running out of plumb – more often than not getting wider – is when cross-joints begin to get noticeably bigger throughout the height of the pier.

lateral pressure on the insides of the pier, sufficient time should be allowed for the mortar to harden before pouring the concrete.

SIMPLE COPINGS TO BOUNDARY WALLS

There are three reasons for providing a distinct finish to the top of boundary walls: first, to stabilize the top of the wall and improve its structural integrity; second, to protect the top of the wall from the ravages of weather; and third, to provide a decorative and aesthetically pleasing finish.

The materials used for finishing off the top of the wall must, therefore, be strong, weather-resistant and reasonably pleasing to the eye. A wide variety of materials can be used to finish off the tops of walls and piers, including brick, pre-cast concrete, stone, reconstituted stone, slate, and so on, but brick (including its use in simple combination with other materials) and pre-cast concrete will be the main materials for discussion here (see Fig 207).

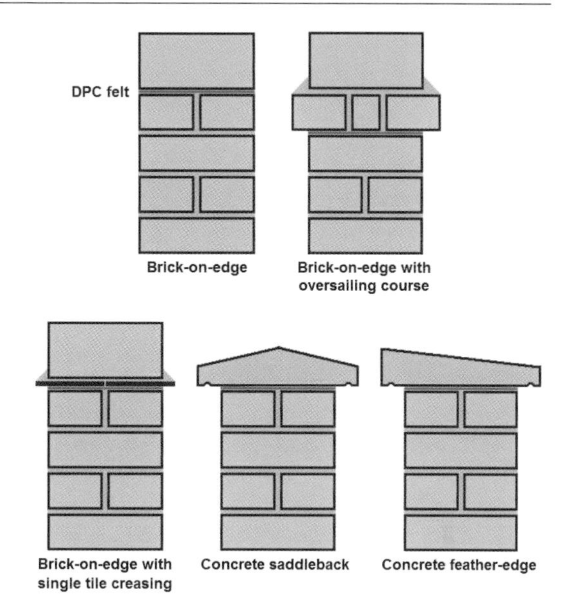

Fig. 207 Simple copings to boundary walls.

Simple Brick-on-Edge Copings

The simplest brick finish to the top of a wall is a brick-on-edge coping. As the name suggests, the wall is finished off at the top with a course of bricks laid on their edge, across the thickness of the wall, with their stretcher face uppermost. When laid in such an exposed situation, the bricks used must be capable of offering a degree of weather protection to the top of the wall and must, themselves, be weather- and frost-resistant. Accordingly, facing bricks, particularly those of ordinary quality, make a poor choice for a brick-on-edge coping and will quickly succumb to frost damage over very few winters. Facing bricks of special quality or engineering bricks should be used.

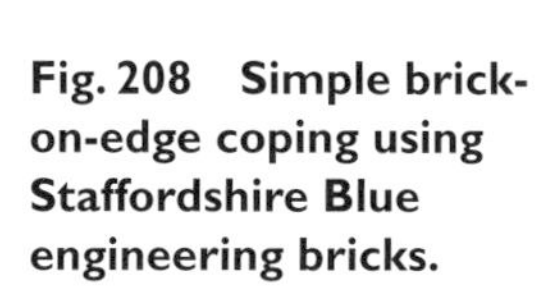

Fig. 208 Simple brick-on-edge coping using Staffordshire Blue engineering bricks.

Fig. 209 Simple brick-on-edge coping: the end brick is held in place with galvanized frame cramps.

Fig. 210 Three bricks are laid at each end and a string line attached.

On a simple brick-on-edge coping, it is important to consider the long-term security of the end bricks of the coping where they are exposed at a stopped-end. Exposed end bricks-on-edge tend to be susceptible to impact and to being knocked off. When possible, it is considered good practice to build in to the bed joint a couple of galvanized fishtail frame cramps.

Laying a Simple Brick-on-Edge Coping

As with all brickwork, the principle of setting up at each end and running-in in between applies equally to brick-on-edge copings.

Start with a minimum of three bricks at each end – any fewer than three will not have sufficient mass (back weight) or stability to withstand the tension of a string-line – make sure that the first brick is plumb, and level across the width of the wall. It should have a 10mm bed joint under it and sit square on top of the wall. This can be checked using a small wooden square.

When laying the second and third bricks, the first brick must be held in place with one hand while the other bricks are squeezed or 'wiggled' into position with the other hand. This is one of the rare occasions when the trowel is put down! Adjusting the bricks into position in this way rather than tapping them with the trowel reduces the chance of disturbing the first brick or bricks, maintains adhesion between the bricks and mortar, and eliminates any damage to the bricks that could be caused with the trowel blade. Cross-joints must be full and solid for the coping to be effective against rain penetration, but excessive mortar should be avoided when applying cross-joints as it makes the bricks difficult to 'wiggle' into position! A measuring tape or a gauge lath held horizontally will help maintain correct and consistent cross-joint thickness.

Once the second and third bricks have been laid, it is important to make sure that they are level across the width of the wall and sit square on top of the wall, and that all three bricks together are level and in line along the length of the wall. Using a spirit level and straight-edge, check that the blocks of three bricks-on-edge at each end are level with each other, then attach a string-line (*see* Fig 210). The dry brick stood on end assists with the attaching of the corner block. On short walls, of course, line and level in between the two ends can be achieved with a spirit level provided the level is long enough

Setting-Out a Brick-on-Edge Coping

Three bricks laid on edge are equivalent in length to one brick, so a wall that is twenty bricks long, for example, will require sixty bricks laid on edge, with 10mm cross-joints, to form the coping.

Fig. 211 The middle of the brick-on-edge coping is set-out by dry bonding.

Where a wall does not 'work bricks', an allowance must be made within the brick-on-edge coping to overcome this problem. Cut bricks within a brick-on-edge coping are not acceptable under any circumstances. The only remaining viable alternative is to make adjustments to the thickness of the cross-joints, either by thickening them to 'lose' the excess in the wall length or by closing them up, in order to fit in an extra brick-on-edge. Knowing which course of action to follow is best achieved by dry bonding the middle section of the brick-on-edge, in order to ascertain how much adjustment is required and in which direction (*see* Fig 211).

Cross-joints should be of a consistent thickness, within a tolerance of + or – 3mm. Adjustment is fairly easy on long walls as very small individual adjustments can be made over the very large number of cross-joints, and these will be hardly noticeable in the context of the whole wall. Short walls with very few cross-joints offer much less scope for adjustment. Very large or very tight cross-joints will result, both of which will look poor. On very short walls that do not 'work bricks', consideration should be given as to whether a brick-on-edge coping is appropriate at all!

Running-In a Brick-on-Edge Coping

When laying bricks on edge to a string-line, there are a number of issues to consider. As is the case with all running-in to a line, it is important to make sure that the brick is laid as close to the line as possible, without touching it. The brick-on-edge coping must be level across its width and, while the string-line provides a reference for line and level at the front, it is very easy to run out of level at the back when attempting to make such a judgement by eye. Accordingly, a method must be found that ensures level from front to back; this can be achieved in one of two ways. The first option is to use a boat level to check every brick for level after it has been laid to the line (*see* Fig 212). Some bricklayers consider this too time-consuming and go for the second option, which is to erect a second string-line at the back to provide level only along the back edge (always remembering that bricks vary in length so this second line is for level only and not for lining in the back edge!)

Variation in Brick Length

It is important to select bricks for a brick-on-edge coping that are reasonably consistent in length. Too much variation is undesirable because the front will be nice and straight from being run-in to a string-line but the back edge/face will be very uneven.

Fig. 212 Each brick-on-edge is laid to line and checked for level front to back.

The disadvantage of two string-lines is that the presence of the second line interferes with the laying process and not all bricklayers have a second set of lines and corner blocks in their tool box anyway.

When laying a brick-on-edge coping to a line, only the smallest dimension of the brick (65mm) is being aligned with the string. This makes it easy and quite common for the brick to lean over from side to side (making the vertical cross-joint taper and become 'V'-shaped). It may also result in the brick not sitting square with the front to the wall (making the cross-joints on top appear tapered and 'V'-shaped). While trying to cultivate a good eye for these issues, it is a good idea periodically to use a boat level to plumb the bricks and a small square to check that the bricks are perpendicular (square) to the main wall line.

Fig. 213 Excess mortar is trimmed from the cross-joints using the point of the trowel.

Finally, it is vital to ensure, as with all facing brickwork, that the bricks are kept clean and free of mortar staining. This can sometimes be a problem on the top surface of a brick-on-edge coping when trimming off mortar from the cross-joints, as using the edge of the trowel tends to drag the excess mortar across the brick surface. Instead, the point of the trowel should be used, and run along the joint.

Laying the last brick in the coping can also prove problematic in terms of keeping the brickwork clean. When laying the last brick-on-edge, place a thin cross-joint on both bricks that have already been laid and thin cross-joints on both sides of the last brick being laid. Gently ease the last brick downwards into place, ensuring that the mortar does not squeeze out too much and smudge the face of the bricks.

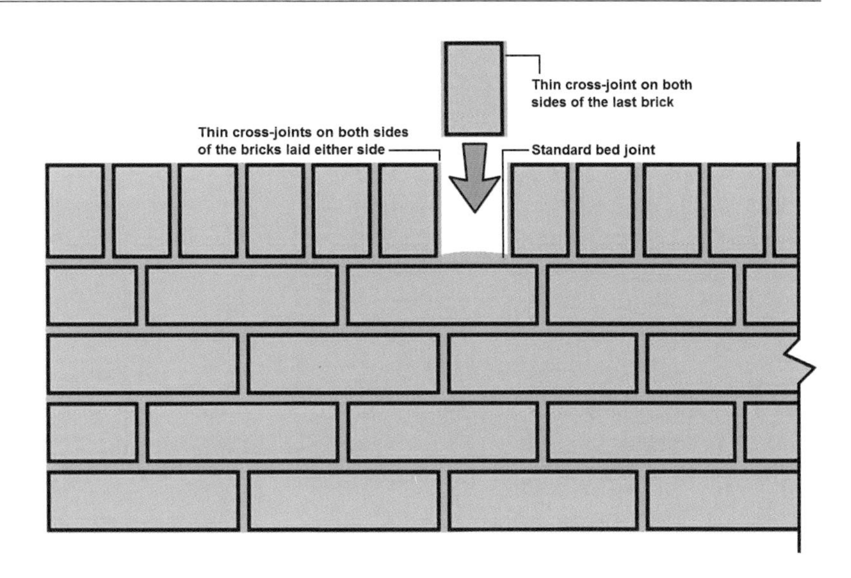

Fig. 214 The last brick-on-edge is laid using a particular method.

LEFT: **Fig. 215 The simplest method of returning a brick-on-edge coping at a corner.**

BOTTOM LEFT: **Fig. 216 Mitred brick-on-edge coping at a corner.**

Simple Brick-on-Edge Copings at Return Angles

When a boundary wall returns at a 90-degree angle or corner, there are two distinct methods of returning the brick-on-edge. The simpler method is also the most common, but there is also a more complex approach that involves mitring the bricks at a 45-degree angle (*see* Fig 216). For the bricklaying beginner, the simpler method may be preferable.

Oversailing Courses

A simple brick-on-edge coping, regardless of how weather-resistant the bricks are, still has one fundamental disadvantage. The front and back of the brick-on-edge coping are flush with the front and back of the wall, with no overhang or projection. This lack of a 'weathering' means that rainwater can simply run down the front and back of the coping on to the face of the brickwork below, leaving it susceptible to frost damage in winter. Unless the main wall has been built from hard-burnt engineering bricks as well, such a finish is best avoided.

One easy way of providing a weathering that will protect the brickwork below the brick-on-edge coping is to introduce an oversailing course (see Fig 218).

The weathering is provided by the oversailing course, which projects up to 30mm from the face of the wall and is finished off with a sloping mortar fillet. The fillet, combined with the projecting brickwork, provides an inclined surface to help throw water clear of the wall face. As much consideration should be given to the choice of brick for the oversailing course as for the brick-on-edge; it is not uncommon to see both constructed with engineering bricks.

Constructing an Oversailing Course

An oversailing course needs to project by no more than 30mm. If perforated bricks are being used, the size and position of the holes in the brick will influence how far the oversailing course is projected – it is not desirable to be able to see the holes when looking under the oversailing course. This is more

Fig. 217 Frost-damaged brickwork below a simple brick-on-edge coping.

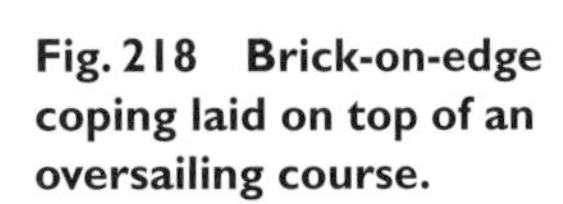

Fig. 218 Brick-on-edge coping laid on top of an oversailing course.

Fig. 219 Gauging a consistent overhang with a notched piece of timber.

of an issue on higher walls, where the top of the wall is above head height.

When laying an oversailing course, the basic principles of setting up at each end and running-in in between remains the same; however there a few more issues to bear in mind. When setting up at each end, the projection must be the same at each end and consistent along the length of each brick. This can be checked with a tape measure, but a quicker and more accurate method is to make a simple gauge by notching a small piece of timber to a depth equal to the size of the overhang.

Next, bricks that overhang have a tendency to tip forward simply because they are being bedded on thin air at their outer edge. To overcome this tendency, the mortar for the bed joint should be spread so that it is higher at the outer edge. The oversailing bricks should be bedded on it so that when they are initially laid, they tip the other way in an almost exaggerated fashion (*see* Figs 220 and 221). The brick can then be adjusted so that it is level across its width. Having started from an exaggerated position in the opposite direction it is much more likely to stay in its final desired position when adjusted, rather than tipping forward. In other words, the natural tendency of the oversailing brick to tip over is being used to the bricklayer's advantage.

Oversailing bricks must always be checked for level across their width, even when being laid to a line. The fact that they project from the face of the

Fig. 220 Exaggerated spreading at outer edge of bed joint for laying oversailing bricks.

Fig. 221 Oversailing brick initially laid out of level, prior to adjustment.

Fig. 222 All oversailing bricks must be checked for level across their width.

wall means that there is no lower reference point to which to flush up the bottom edge of the oversailing brick (see Fig 222).

The height of the wall also dictates the way in which an oversailing course is run-in; specifically, whether it is run-in to the top edge or the bottom edge. Where projecting brickwork is concerned, the eye tends to be drawn to the nearest, prominent edge. On a high wall, the observer's eye will be drawn to the bottom edge, which is the nearest. If the top edge of the oversailing course on a high wall is run-in to a line, any slight deviation in brick thickness will show as an undulating deviation in that bottom edge. On this basis, oversailing courses on high walls tend to be run-in to the bottom edge of the bricks.

Oversailing Courses at Stopped-Ends of Walls

For boundary walls where there is no attached pier at the end of the wall (or where the pier has

Fig. 223 An oversailing course can be extended to provide a weathering at a stopped-end.

Brick-on-Edge Copings with an Oversailing Course

Whether there is an oversailing course present or not, the method adopted to con-struct a brick-on-edge coping is the same. Where there is an oversailing course, how-ever, the brick-layer must make sure that the brick-on-edge is set back from the edge of the oversailing course a distance equal to the projection of the oversailing course. In other words, he or she must ensure that the brick-on-edge coping effectively sits on top of the main wall. This can be achieved by simply turning over the home-made wooden gauge and using it to measure how far to set the brick-on-edge back from the front face of the oversailing course.

been terminated below the top of the wall), consideration must be given to extending the oversailing course beyond the end of the wall, in order to provide a protective weathering to the stopped-end brickwork.

If the oversailing course is finished flush with the stopped-end, the stopped-end brickwork is left vulnerable to the vagaries of the weather. Good practice, therefore, dictates that the oversailing course should be made to overhang beyond the stopped-end of a wall in order to provide a weathering (*see* Fig 233). An overhang at the end, however, does mean that the length of the oversailing course will exceed the length of the original wall. As a result, cuts or a broken bond will need to be introduced in the middle of the oversailing course so that the bond is maintained. Also, where any oversailing course is visible at the stopped-end of the wall, a cut must be introduced at the end to close the visible gap between the front and back oversailers (*see* Fig 207). The enlarged collar joint along the wall length between the front and back of the oversailing course is commonly filled with mortar, but it is good practice to in-fill with cut bricks.

Tile Creasing

As an alternative to a brick oversailing course, one or two courses of creasing tiles can provide the overhang, with the weathering completed with a mortar fillet (*see* Fig 207). Creasing tiles are rectangular, plain, flat, clay tiles that measure 265 × 165 × 11mm, available in a number of colours such as brown, red and blue brindle. The size and flat nature of the tiles allows a larger overhang (up to 40mm) to be created, which ensures a better weathering to the wall.

On both single- and double-course tile creasings (*see* Figs 225 and 226), all bed joints are 10mm and

Fig. 224 Red creasing tiles.

Fig. 225 Single-course tile creasing.

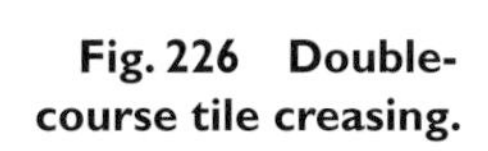

Fig. 226 Double-course tile creasing.

Fig. 227 Nibbed clay roofing tiles used as a single-course tile creasing.

cross-joints should be as tight as possible and no more than 3mm. Where a double course of tiles has been used, the tiles should be overlapped or half-bonded, for structural strength and water resistance. The methods of laying, levelling and running-in courses of creasing tiles are exactly the same as those used for brickwork.

Some bricklayers take the view that the straight finish of creasing tiles is rather dull. The decorative effect may be enhanced by using clay roofing tiles with nibs as a single course under the brick-on-edge coping.

If required, it is also possible to create a double-course creasing by running a course of plain creasing tiles, in a matching colour, on top of the course of nibbed tiles. If using clay roofing tiles in this way, it is wise to select tiles that do not have too much of a surface camber, as this can make laying and levelling rather difficult.

It is good practice with a tile creasing, as with an oversailing course, to extend it at a stopped-end in order to provide a weathering and protection to the stopped-end brickwork.

Mortar Fillet

To complete the weathering, the projecting brickwork or creasing tiles must be provided with a sloping finish to shed rainwater. This is achieved with a mortar fillet (see Fig 207), which should be a flat, sloping surface of between 35 and 45 degrees.

To ensure that the top edge of the fillet runs to an even, straight line, attaching a string-line is much quicker and more accurate than trying to do it by eye.

Load the brick trowel blade with mortar and, using the back of a pointing trowel, scrape mortar from the brick trowel and 'rough-in' the fillet to the line. Use the pointing trowel in a downward

TOP: **Fig. 228 Nibbed clay roofing tiles used as the lower course of a double-course tile creasing.**

MIDDLE: **Fig. 229 Tile creasing extended to provide a weathering at a stopped-end.**

BOTTOM: **Fig. 230 String-line used as a guide for the top of a mortar fillet.**

Fig. 231 Mortar fillet 'roughed-in' to the string-line.

motion from the line, using the edge of the over-sailing course (or tile creasing) to 'clean' the back of the pointing trowel. Try to keep the diagonal, downward movement of the pointing trowel as straight as possible, to achieve a flat surface to the sloping fillet. It is very easy to end up with a convex curve to the fillet unless care is taken at these early stages.

Remove the string-line and, if required, apply small amounts of mortar with the pointing trowel to fill in any obvious dips or hollows in the fillet. Leave the mortar to go off, as is the practice prior to jointing-up brickwork. Then, using the back of a pointing trowel, polish and compact the surface of the fillet by drawing the trowel along the length of the fillet. This process is vital to ensure a smooth finish and the weather resistance of the mortar fillet and is greatly assisted by keeping the pointing trowel wet. Have a bucket of water to hand for this purpose.

Fig. 232 Mortar fillet polished and compacted with a wet pointing trowel.

Fig. 233 Finished mortar fillet.

Fig. 234 Pier terminated below the top of an old wall using plinth bricks to form a 'tumbling in'.

PIER CAPS

When terminating the top of an attached pier below the top of the main wall, it is necessary to provide a weathering to the top. This is usually achieved by way of special bricks called 'plinths' (see Fig 234). The inclined surface of the plinth bricks, which are available as headers and stretchers, provides a slope to shed rainwater. At the same time, it maintains the bond of the pier as it gradually reduces in width until the face of the main wall is reached and the pier terminated. Terminating a pier in this way is known as 'tumbling in'.

Attached piers can be terminated at the same height as the boundary wall, meaning that the finishing or coping to the top of the pier is integrated with the coping on top of the wall (see Fig 235). This approach is the least common and tends to be found on comparatively low boundary walls.

It is far more common for the finished height of an attached pier to exceed that of the wall, emphasizing the pier as a decorative feature (see Fig 236). As this method is the most favoured, subsequent advice given here on pier caps and copings to boundary walls will relate to that method.

Fig. 235 Attached pier terminated at the same height as the wall.

Fig. 236 Attached pier terminated higher than the wall.

Attached piers that terminate above the height of the main wall can be capped in exactly the same way as isolated piers. On attached piers, the method of capping should, ideally, match the coping used for the main wall in terms of design and choice of materials. The constructional methods detailed above for boundary walls can equally be utilized on top of piers. However, the most common approach will include a weathering to all four sides, formed either with a brick oversailing course (see Fig 237) or a tile creasing (either single- or double-course) but both incorporating a mortar fillet. Where any mortar fillet returns round a corner or angle, the fillet must be finished with a neat, sharp mitre at the apex of the corner. A small cut or closer is also required to be incorporated into a brick oversailing course to maintain the bond. Such a closer will appear in the same relative position on opposite sides of the pier, with its size determined by the dimensions of the pier and also the extent of the overhang.

For strength and aesthetics, any brick-on-edge pier capping must be bonded 'on plan' (see Fig

Fig. 237 Bonded brick-on-edge pier cap.

Brick-on-Edge Pier Caps

Piers, whether isolated or attached, should be finished off with a brick-on-edge capping only where the pier's width works to full brick dimensions. For example, a pier that measures 1½ x 1½ bricks square cannot accommodate five bricks laid on edge as the cross-joints will be too tight. Four bricks laid on edge will result in cross-joints that are too wide. Accordingly, piers should be sized carefully to work full bricks in at least one of its dimensions if a brick-on-edge capping is required.

237). The presence of a weathering and mortar fillet provides an opportunity for the cut face of the half-bricks to be invisibly accommodated in the bed joint beneath the coping so that no fair-faced cutting is required.

PRE-CAST CONCRETE COPINGS AND CAPS

Pre-cast concrete units (see Fig 207) are a cheaper alternative to a brick-on-edge finish, being much quicker to construct. They have other advantages in that they are designed to be wider than the wall or pier so have the weathering or overhang in-built. They are commonly cast with a sloping surface to shed water and a groove or throating is often cast into the underside edge. This acts as a drip to ensure that rainwater that tracks under the coping falls clear of the wall face below. As they are large in size (typically 600mm long), the number of cross-joints in the overall length of the wall is greatly reduced. This is advantageous as the cross-joints are potentially vulnerable points where rain-water can penetrate, particularly if they are inadequately filled. The disadvantages are that pre-cast copings do look rather utilitarian or industrial, and pre-cast units for piers are generally only available for square piers, so the choice is somewhat limiting.

The two most common pre-cast concrete copings for walls are 'feather-edge' and 'saddle-back'. Saddleback pier caps are also available.

When laying pre-cast copings, the same principles apply as when laying brick-on-edge copings. Lay one coping at each end, level with one another and on 10mm beds of mortar, then run in to a string-line. Before running in, ensure that the overhang front and back is equal and consistent along the length of the coping – check it with a measur-

TOP: **Fig. 238 Pre-cast concrete feather-edge coping.**

CENTRE: **Fig. 239 Pre-cast concrete saddleback coping.**

BOTTOM: **Fig. 240 Pre-cast concrete saddleback pier cap.**

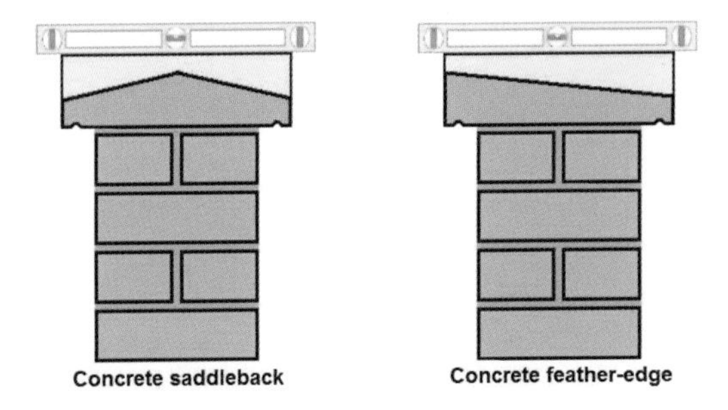

Fig. 241 Timber profiles to assist widthwise levelling of concrete copings.

ing tape to ensure accuracy. One key consideration is how to set the end copings level on top of the wall both widthwise and lengthwise. In terms of levelling along the length, a spirit level can be placed on the underside of the overhanging edge. However, this is not completely reliable, as the underside surface of the concrete cannot be guaranteed to be flat or even. Instead, level the coping along the top, making sure that the spirit level is placed exactly parallel with the edge of the coping. Failure to do this will provide a false reading due to the spirit level being on a slope. Levelling the width is a little trickier and the only completely accurate method is to manufacture a timber template that will accommodate the profile shape of the coping and provide a flat surface on which to place the spirit level (*see* Fig 241). The cutting line for the profile is marked simply by using the top of one of the copings as a template.

As with a brick-on-edge coping, concrete copings can be run-in using two string-lines – one at the front and one at the back – or by using one string-line and levelling each coping across the width. As is the case with brick oversailing courses, if the concrete coping is positioned above head height, the string-line should be positioned so as to run-in the copings to the lower edge, as this is the edge to which the eye-line will be drawn.

Cross-joints between copings should not exceed 10mm but it is common for joints to be made tighter (6mm) to increase weather resistance. Bed joints under concrete copings should be jointed-up as should cross-joints, which should be jointed all the way round, including under the overhang. The groove or throating under the edge of the coping must be continuous along the length of the wall, so the mortar at cross-joints should be jointed flat into the throating with a flat-bladed jointing iron.

Good practice with concrete copings, as with an oversailing course or a tile creasing, is to extend it at a stopped-end in order to provide a weathering and protection to the stopped-end brickwork.

Fig. 242 Concrete coping extended to provide a weathering at a stopped-end.

Setting-Out and Cutting Copings

More often than not, copings will need to be cut in order to fit the length of the wall. Any cuts should be placed in the middle of the wall where the coping has at least one free end. Where the coping is laid in between two attached piers, cuts of equal size are usually placed at both ends, adjacent to the piers and not in the middle. Small cuts should always be avoided as they are very noticeable. It is better to have two or even three cuts of larger size that blend in better than one small, highly visible cut. Where the wall on which the copings will be installed incorporates a vertical expansion joint (*see* Fig 244), the copings must be set-out and installed in such a way that allows the expansion joint, and the flexible materials contained therein, to continue through the coping as well. The copings should be set out dry on top of the wall in order to establish the size and number of any cuts. Cutting copings cannot be achieved by hand so a mechanical disc cutter with a diamond blade should be used.

DAMP PROOF COURSES IN BOUNDARY WALLS

While it is not required for the purpose of any statutory obligation, it is considered good practice to include a horizontal damp proof course at the base of a boundary wall and/or isolated pier. The inclusion of a DPC, typically positioned around 150mm above finished ground level, prevents groundwater from rising that would ordinarily cause staining to the facing brickwork and/or efflorescence from soluble salts contained in the groundwater. In addition, frost damage may occur in winter to the brickwork at the base of the wall resulting from groundwater in the brickwork expanding on freezing and causing the bricks and/or mortar to crack and spall. For boundary walls, a layer of DPC felt incorporated into a bed joint 100–150mm above finished ground level, or two courses of solid Staffordshire Blue engineering bricks, will provide a sufficient barrier to prevent the rise of groundwater. For more detail on materials for DPCs and their installation, *see* Chapter 11.

An ideal boundary wall construction (*see* Fig 243) includes all of the key features required by

Fig. 243 Finished boundary wall.

good practice. Specifically, and working upwards from ground level, the masonry below ground level will have been constructed of common bricks or concrete blocks of a quality designated as suitable for use below ground. The wall has a two-course solid blue-brick DPC, which is an ideal moisture barrier and also forms an attractive feature.

Brickwork from ground level up to DPC is constructed in engineering bricks of special quality that are also water- and frost-resistant, making them capable of remaining in a saturated condition (either from groundwater or rainwater 'splash-up' from the pavement) with little risk of frost damage in winter. Engineering bricks are also more resistant to the dirt and staining that are associated with rainwater splash-up. In addition, the low water absorbency of engineering bricks will minimize the possibility of efflorescence – ordinarily, any soluble salts present in the ground will be absorbed during periods of wet weather, resulting in white salt deposits forming on the face of lower-quality bricks during dry weather.

The wall incorporates a plentiful number of double attached piers for maximum stability, which terminate higher than the main wall. This is both attractive and further increases stability. The brick-on-edge copings and caps to the wall and attached piers respectively are constructed in Staffordshire Blue engineering bricks for maximum weather resistance. Oversailing courses with a mortar fillet have been included to the tops of the wall and all piers to provide a weathering under the brick-on-edge coping. Closer inspection of the brickwork shows that a half-round joint finish has been applied; this is one of the more compact and weather-resistant joint finishes available.

EXPANSION JOINTS IN LONG BOUNDARY WALLS

Masonry structures expand, contract, lengthen and shorten due to changes in temperature, moisture content of the ground and relative humidity of the surrounding air. As a rule of thumb, brick walling expands and contracts by approximately 1mm in every 1m length of walling so the longer a wall is, the greater the extent of the movement overall. This is less of a problem for most domestic construction but is more significant on very long free-standing walls, which can be many metres in length.

Provision needs to be made in such walls to allow for this structural movement, usually in the form of an expansion joint (*see* Fig 244), which effectively acts as a 'separating point' between two adjacent lengths of walling. A continuous, straight, vertical 12mm joint is introduced, and filled with a strip of semi-rigid fibre-board, instead of continuing the bonding arrangement through. It is good practice to introduce an expansion joint for every 6m length of walling. The width of the expansion joint, in this instance 12mm, is derived from two adjacent walls of 6m both having the capacity to

> **DPC Felt under Copings**
>
> In order to provide complete protection to the top of a boundary wall (or pier), it is considered good practice to incorporate a layer of DPC felt, as shown in Fig 207. There are no hard and fast rules on this point and a view should be taken on whether it is actually necessary, based on the choice of materials for the coping or cap, the weather resistance of those materials, and the likely level of weather exposure to which the wall will be subjected.

Fig. 244 Expansion joint with slip tie in situ.
Ancon Building Products

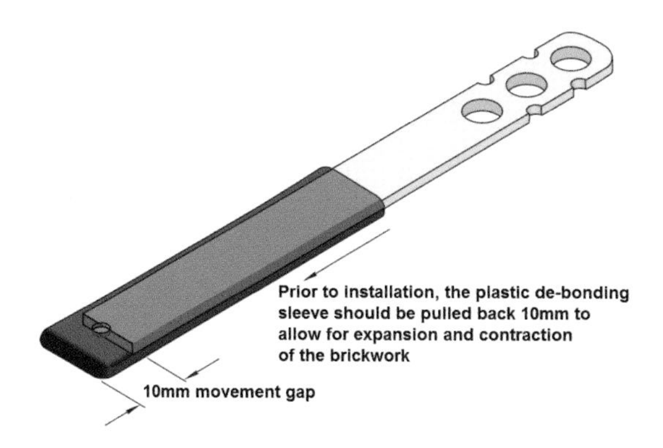

Fig. 245 Expansion joint slip tie.
Ancon Building Products

expand/contract 6mm each (in other words, 1mm for every linear metre). Expansion joints are usually only installed above DPC level, and not below, but must continue through any coping at the top of the wall.

Structural stability across expansion joints is provided by special metal de-bonded wall ties, placed every three courses. These are known as 'slip ties', since they tie the two sections of walling together across the expansion joint, but one end of the tie is fitted with a moveable plastic sleeve that acts as a 'de-bonding element' between the mortar joint and the metal tie. This allows the walling to expand and contract – the tie is rigidly fixed at one end but at the other only the plastic sleeve and not the metal part of the tie is gripped by the mortar bed joint. The plastic part of the tie remains static while the metal section 'slips' within it during expansion and contraction.

Despite the presence of expansion joints effectively dividing the wall into separate sections, the wall is still constructed as if it were one continuous length, with the two extreme ends set-up and the middle run-in to a string-line. As the wall is run-in, provision must be made to allow for the inclusion of the vertical expansion joint as the work proceeds and great care must be taken to ensure that the expansion joint is truly vertical, or plumb. The most accurate method is to 'build in' a straight 1200mm long timber lath at the front of the expansion joint (see Fig 246), which acts as a

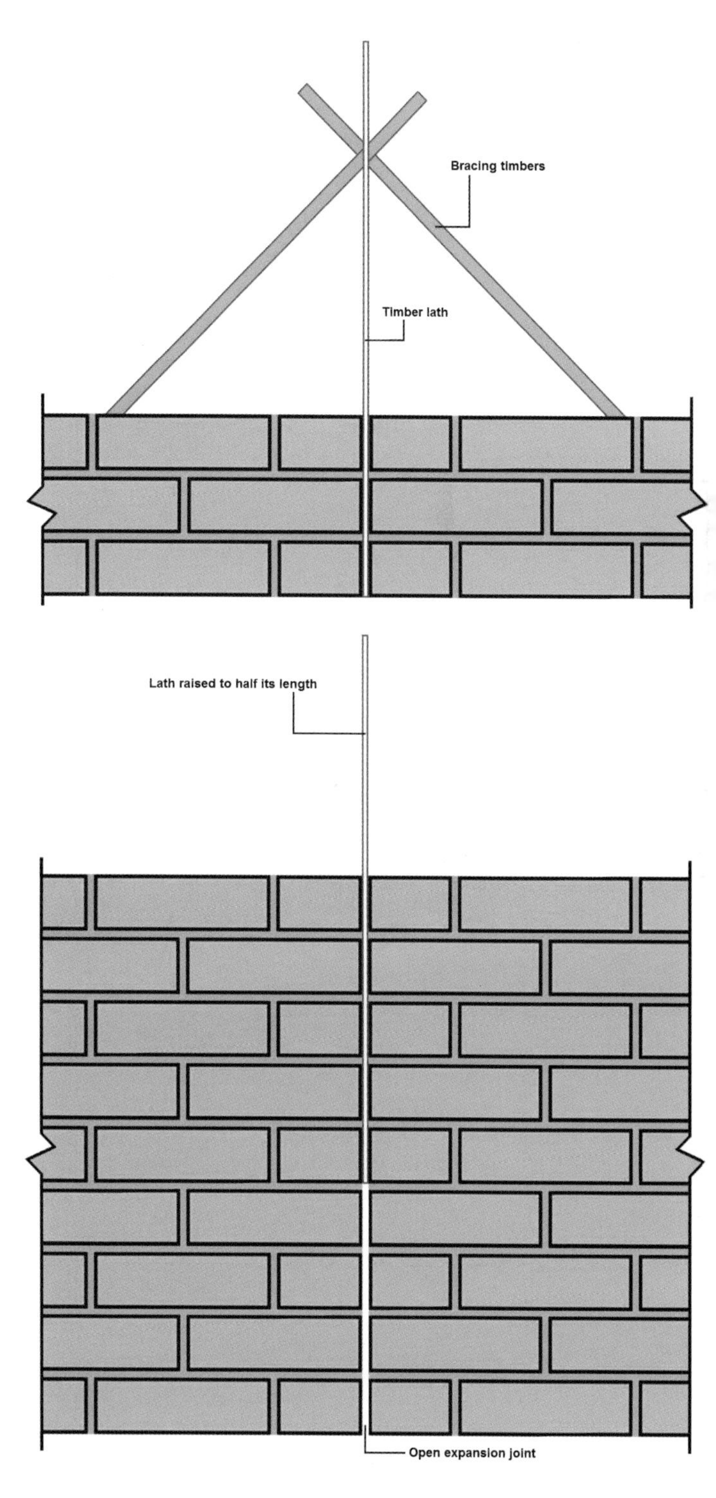

Fig. 246 Method of forming an open expansion joint.

profile to form an open expansion joint. The lath must be equal in thickness to the width of the expansion joint but not the full depth of the brick skin, as this would prevent the installation of the slip-ties across the expansion joint. The timber lath should be positioned vertically and, initially, braced in position. (This temporary bracing should be set behind the brickwork so that it does not interfere with running-in the wall to a string-line.) As the wall increases in height to half-way up the lath, the temporary bracing can be removed, as the brickwork either side will hold the lath in place. The lath should be periodically checked for plumb to maintain the accuracy of the expansion joint. When the brickwork reaches the top of the lath, the lath is raised vertically by about half its length so that it remains plumb and gripped in position. To make it easier to raise the lath and, eventually, to remove it completely, it is a good idea to tap it on completion of every four or so courses to loosen the mortar bond inside the expansion joint.

Once the wall has been completed, the timber lath is removed and the open expansion joint is thoroughly cleared of any mortar droppings and debris. The fibre-board strip is then inserted into the open joint. Slits will need to be made where slip-ties occur or separate short lengths of board could be used that fit vertically in between the ties. The fibre-board that is used to fill the expansion joint during construction is not at all weather-resistant and will rot away over time. To protect it and also to provide a neat finish to the expansion joint, the exposed edge is completely sealed with flexible mastic (in a colour to tone or match with the brickwork). This means that the fibre-board needs to be set back 5–10mm from the face of the wall as it is installed. Strips of polyethylene offer a weather-resistant alternative to fibre-board, although the joint will still need to be sealed with mastic, as an aesthetic consideration.

Fig. 247 Finished expansion joint sealed with mastic.

CHAPTER 13

Simple Decorative Work

DECORATIVE BRICKWORK

Despite an individual brick being nothing more than a simple cuboid, bricks may be combined together in many different ways in order to create visually stunning decorative patterns and shapes of enormous complexity. These decorative applications can be greatly enhanced by the myriad of brick colours and textures that are available. The results certainly give some credence to the notion that a bricklayer is not just a craftsman, but an 'artist in burnt clay'. The aim here is to provide some ideas for enhancing the appearance of the type of brickwork already described, using a number of simple decorative applications. In addition, there is a vast array of brickwork in the built environment, which can give additional inspiration.

By its very nature, decorative brickwork is intended to draw attention to visual features in the walling and to be a focal point in itself. With this in mind, it is vital to consider the following key points:

- All decorative work must be truly level and plumb.
- The angles at which bricks are laid must be carefully checked.
- Any projecting or oversailing brickwork must be carefully lined in to the edge or arris to which the eye-line will be drawn. The general rule, for example, is that work at a high level, above head height, should be lined in to the bottom edge, and the reverse is the case for work at a low level. There are some exceptions to this rule and these will be highlighted.
- Any cuts should be carefully marked out and cut neatly and accurately.
- Individual bricks for decorative work should be carefully selected and any defective and/or damaged bricks should be rejected.
- Bricks used for decorative features should be carefully selected to be of uniform size. Any variations will be very noticeable and will spoil the overall effect, with over- or undersized bricks resulting in uneven mortar joints.
- The location of decorative features should be carefully chosen. Many such features can interfere with the bonding arrangement of the main wall and, in themselves, sacrifice structural strength in favour of aesthetics. Accordingly, they should not be used in significant load-bearing situations: for example, a decorative panel on an isolated gate pier would be acceptable but not where the pier is supporting the end of a structural beam.
- Concentration is required when building decorative features – there is nothing worse than suddenly remembering that a contrasting-coloured brick should have been inserted six courses ago!

Contrasting Coloured Bricks

Probably the simplest method of introducing a decorative feature is the use of bricks that are a different or contrasting colour from the majority of bricks used in the main wall. This could involve several courses of contrasting bricks made to form a contrasting band, or contrasting bricks used to pick out patterns in the bonding arrangement. In both cases, the main bonding arrangement of the wall has not been altered at all, so its structural strength has not been compromised.

Where a decorative feature projects outwards

Fig. 248 Walling with contrasting bands of different-coloured bricks.

Fig. 249 Decorative pattern picked out in projecting bricks of a contrasting colour.

(*see* Fig 249), creating a ledge on which rainwater may collect, it is wise to use bricks such as Staffordshire Blue engineering bricks, which will not suffer spalling or frost damage as a result. If a softer, contrasting facing brick had been used, such a feature would, almost certainly, have been laid flush with the rest of the wall.

With such a huge variety of colours and textures of brick available, the possibilities for introducing and mixing contrasting bands and patterns are almost endless. The only limiting factor, it would seem, is the imagination of the designer or bricklayer!

Diaper Patterns

'Diaper' or 'diapering' are old heraldic terms that relate to the enlivening of flat, plain surfaces with the application of a variety of shaped patterns. The shapes applied tend to be squares, rectangles or, most commonly, diamond or 'lozenge' shapes, similar to those seen in Fig 249. True diapering, however, tends to be on a much grander scale (*see* Fig 250).

Fig. 250 Diaper patterns.

Fig. 251 Projecting single diaper.

Large-scale diapering can involve simply picking out larger, obvious diamond patterns in the original bonding arrangement, but these tend to look rather horizontally elongated. True diapering tends to favour height over width and requires a degree of cutting and broken bonding in order to achieve the required pattern. In the example in Fig 251, the stretcher bond in the main wall has had to include a lot of cutting in order to accommodate the projecting single diaper pattern.

Dentil Courses

A dentil course is formed by projecting or recessing every alternate brick from the main wall face. Dentils are formed with headers with a projection or recess of a maximum of 28mm, although 18mm to 20mm is the norm. Dentil courses are more commonly formed using projections.

In one common application, a dentil course is sandwiched between two oversailing courses to form a decorative band; this is particularly effective at first-floor level on a house (*see* Fig 253). When constructed in this way, the resultant decorative feature is often referred to as a 'string course' and may be limited to an individual elevation or may extend all the way round the building. When building projecting string courses, the lower eye-line should be formed by lining in the bottom arris of the first oversailing course (as would be expected). The upper eye-line should be formed by lining in the top arris of the second oversailing course, which goes against the usual rule of thumb. Despite being above head height, the top arris is the most prominent edge in this case because the eye-line along

Fig. 252 Dentil course placed immediately beneath an oversailing course – a typical application, at the top of a boundary wall.

Fig. 253 Dentil course within an oversailing band.

Fig. 254 Soldier course in a typical application, above a window opening.

the bottom edge is 'softened' by the presence of the dentil course. Again, the projections have been formed with Staffordshire Blue engineering bricks because of their high resistance to frost.

Projecting headers for a dentil course and other decorative features will lead to a corresponding recess on the back of the wall. This is of no consequence where the back of the wall is not seen. However, where the back is visible, a recess is undesirable, both from an aesthetic point of view and because it provides a ledge for rainwater to sit on. One solution is to cut or 'snap' the headers in half in order that one half can project forward on the front of the wall and the other half can be laid flush at the back. This type of application gives rise to the term 'snap header'.

Building Projecting Features

When running-in courses of brickwork that include any isolated projecting bricks, it is necessary to run-in the main walling first and then lay any projecting bricks last in each course so as not to interfere with the string-line. All projecting features must project uniformly and equally. A purpose-made timber template should be used to gauge projections to ensure accuracy, as recommended during the construction of oversailing courses (*see* Chapter 12). A projecting dentil course can be run-in to a string-line once the two end projecting bricks have been laid and projected correctly.

Soldier Courses

A soldier course is a course of bricks laid vertically upright/on end. It is typically located directly over the top of a door or window opening. The length of the soldier course over an opening is usually the same as the width of the opening between the reveals.

There are a number of important issues to consider when building soldier courses (*see* Fig 255):

- The bricks selected for use as a soldier course must not vary in length otherwise this will reflect in an uneven bottom arris when the top arris is run-in level to a string-line. Even though the bottom arris is often the one that forms the eye-line, it is still usual to lay soldiers to a string-line at the top, hence the importance of ensuring that the bricks are all of the same length.
- The widths of door and window openings do not always 'work bricks' so it is vital to dry bond the soldier course first to see what adjustment is required to the thickness of cross-joints, if any. It is a good idea to use a pencil or chalk to mark the top of the supporting steel lintel with the positions of the cross-joints once the soldier course has been set-out dry.

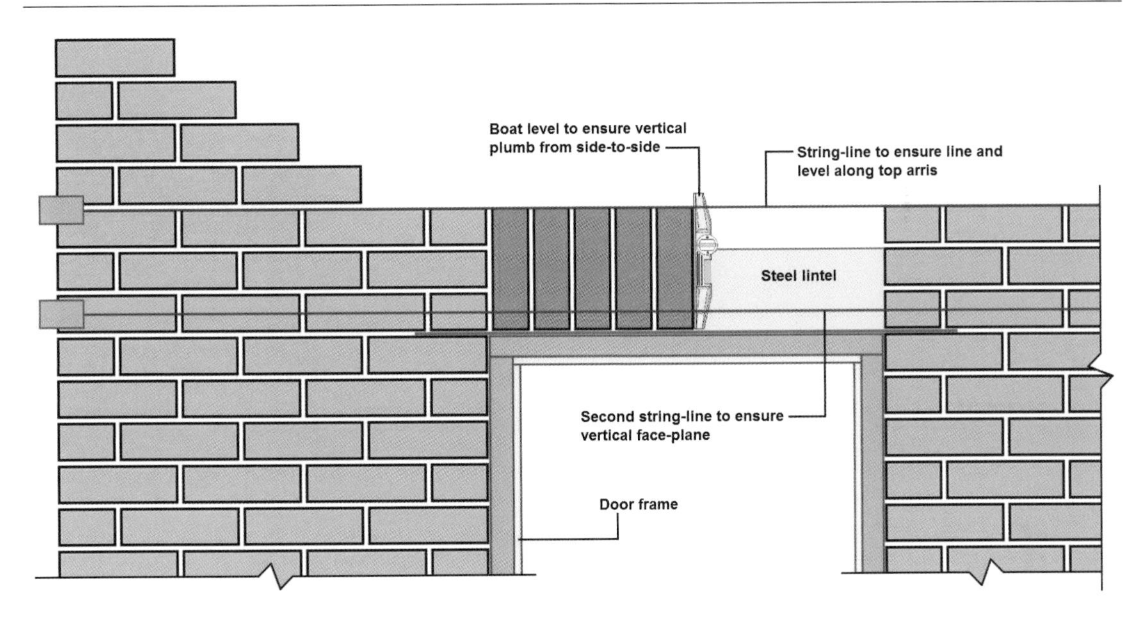

Fig. 255 Method of constructing a soldier course above a door opening.

- Laying a soldier course to a string-line will also prevent the face of the bricks being twisted, which would ruin the face plane. On short soldier courses (for example, over a door) it is usual to lay soldiers from one side straight to the other. Longer soldier courses are usually started from both ends and run-in to the middle.
- In the context of face-plane deviation, the soldier bricks must be vertical up their face and must not 'kick in' or 'kick out' at the bottom. Verticality up the face of the soldier bricks is

Fig. 256 Soldier course below a window opening at sill level.

Fig. 257 Soldier string-course in contrasting coloured bricks.

ensured by using a second string-line near the bottom of the soldier course.

- Every soldier brick must be truly vertical and not lean to one side. Each one must be checked with a boat level as it is laid.
- Finally, when laying a soldier course over a door or window opening, weepholes must be included.

Laying the last soldier brick can also prove problematic in terms of keeping the brickwork clean. When laying the last soldier, thin cross-joints are placed on both bricks that have already been laid and on both sides of the last brick being laid. The last soldier is gently eased downwards into place, ensuring that the mortar does not squeeze out too much and smudge the faces of the bricks.

Soldier courses are not limited to appearing above door and window openings – they are also often seen beneath window openings at sill level.

Soldier courses can be incorporated into a free-standing wall, or can be extended to the whole width of a building elevation or all the way round a building at first-floor level, to form a decorative band or string-course.

SIMPLE DECORATIVE PANELS

The use of a decorative panel is quite a common method of enhancing the appearance of plain brick walling. While a panel can be projecting or recessed in relation to the surrounding brickwork, it is usual for it to be left flush with the face plane of the surrounding wall.

Panels can be incorporated into free-standing walls or buildings. On buildings, it is usual to find them positioned beneath window openings due to

Fig. 258 Basketweave panel under construction within a pre-built square opening.

their lack of structural or load-bearing strength. Regardless of location, the usual starting point is to leave an opening in the brickwork ready to receive the panel once the surrounding brickwork has been built up to the top of the opening/ proposed panel. In the example of a 'basketweave' panel (see Fig 258), the opening that has been left, with allowances for 10mm mortar joints, is three bricks (685mm) wide and nine courses (675mm) high. Once the panel has been built into the opening, the remaining brickwork can be continued straight over the top of the panel. Panels need not be square but, in order to ensure some symmetry to the pattern, it is always best if the width is a multiple of full bricks and the height is a multiple of three courses. When constructing panels, it is vital that the insides of the opening are perfectly plumb and that the brickwork either side of the opening is the same level.

There are many different patterns of brickwork that can be formed into a panel and the same patterns may be reoriented through 45 degrees in order to give a number of variations. However, this approach requires complex setting-out and numerous angled cuts, and it is better to begin with basketweave or diagonal herringbone – two simple patterns with square cuts.

Basketweave Panels

Basketweave consists of groups of three bricks laid alternately vertically and horizontally. Construction begins with an opening formed in the main wall ready to receive the panel; next, the panel bricks are laid in the general order shown (see Fig 259). This is not the only order that can be used but it is the one that will prove the easiest and most convenient for the bricklayer. When installing the bricks into the panel, all stretchers are laid to a string-line as the bed joints will coincide with the bed joints in the main wall. Bricks 1 and 2 will be laid first, the line will be moved up to lay bricks 3 and 4, then moved up again to lay bricks 5 and 6, followed by the three soldiers bricks (7, 8 and 9). The soldier bricks must be laid vertically and each one must be checked up its vertical side with a boat level, and the front face plane, top to bottom, lined in to the surrounding brickwork with a straight-edge or spirit level. This general pattern of working is repeated as the panel is built upwards, then the brickwork of the main wall is continued over the top of the panel.

Diagonal Herringbone Panels

Diagonal herringbone consists of a series of patterns of bricks – stretchers and soldiers laid at 90 degrees to one another, creating a diagonal pattern at 45 degrees to the horizontal (see Fig 260). Whilst the herringbone pattern still comprises only soldiers and stretchers, it is much more complex than basketweave and requires a considerable amount of cutting. The cuts are all square cuts and only bats of 65mm long and 140mm long are required to complete the pattern, but the cuts must be made very accurately and all to the same length, as errors will show up very obviously in the mortar joints. The relative complexity of the pattern also demands great concentration from the bricklayer as it is easy to make a mistake, especially if bricks of two colours are being used.

The panel bricks are laid in the order shown in Fig 260, starting from the bottom right-hand corner. In this case the use of a line for only one stretcher at a time becomes impractical, so stretchers are laid and checked for level with a boat level and the face plane lined in to the surrounding brickwork with a straight-edge or spirit level. The soldier bricks must be laid vertically and each one must also be checked up its vertical side with a boat level, and the front face plane lined in to the surrounding brickwork with a straight-edge or spirit level. This 'up and down from one side to the other' pattern of working is repeated as the panel is built, then the brickwork of the main wall is continued over the top of the panel. Great care must be taken to ensure the correct thickness of mortar joints so that they line up properly with the bed joints of the brickwork in the main wall. An overly thin bed joint under one soldier brick can very quickly disturb the symmetry of herringbone.

SIMPLE STRAPWORK ON PILLARS

Decorative panels can also be incorporated into the face of an isolated or attached pier, with the same principle applying of leaving an opening in the pier into which the panel is inserted. In order to have a meaningful impact, the panel needs to

Leave a recess into which to build the panel

Construct the panel brickwork in the order shown

18
10 11 12 17 15 14 13
16
5 6
3 7 9 8 4
1 2

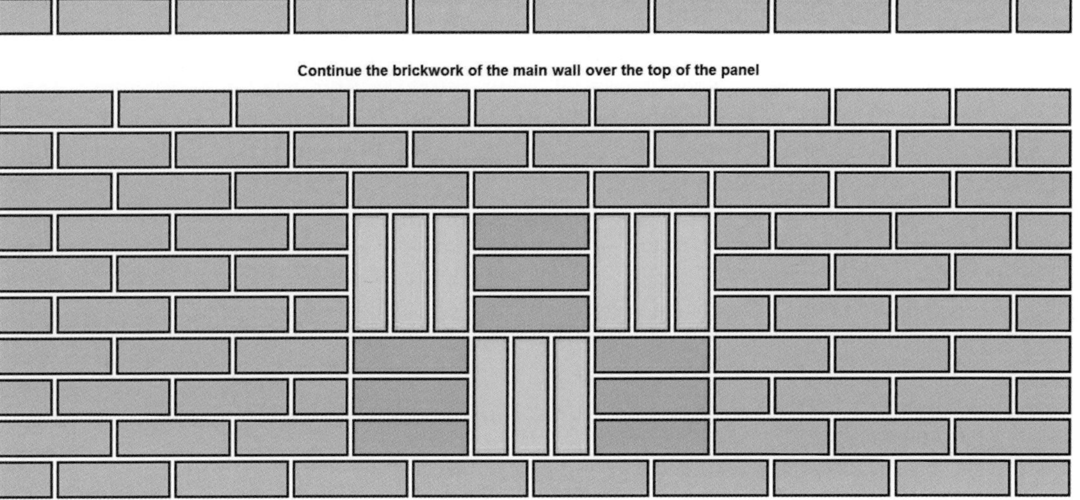

Fig. 259 The general order of the method of construction of a basketweave panel.

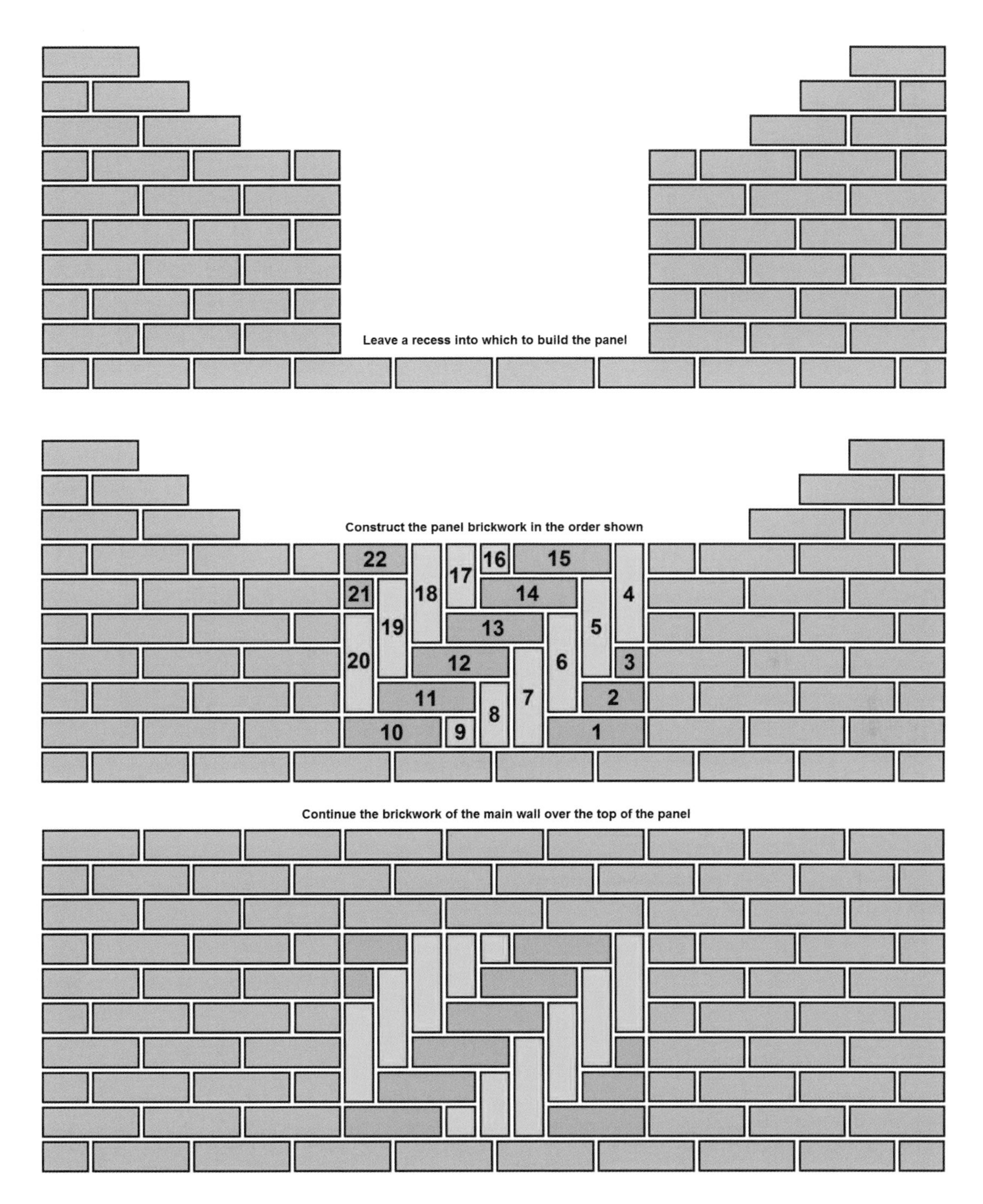

Fig. 260 Method of constructing a herringbone panel, again starting with an opening formed in the main wall ready to receive the panel.

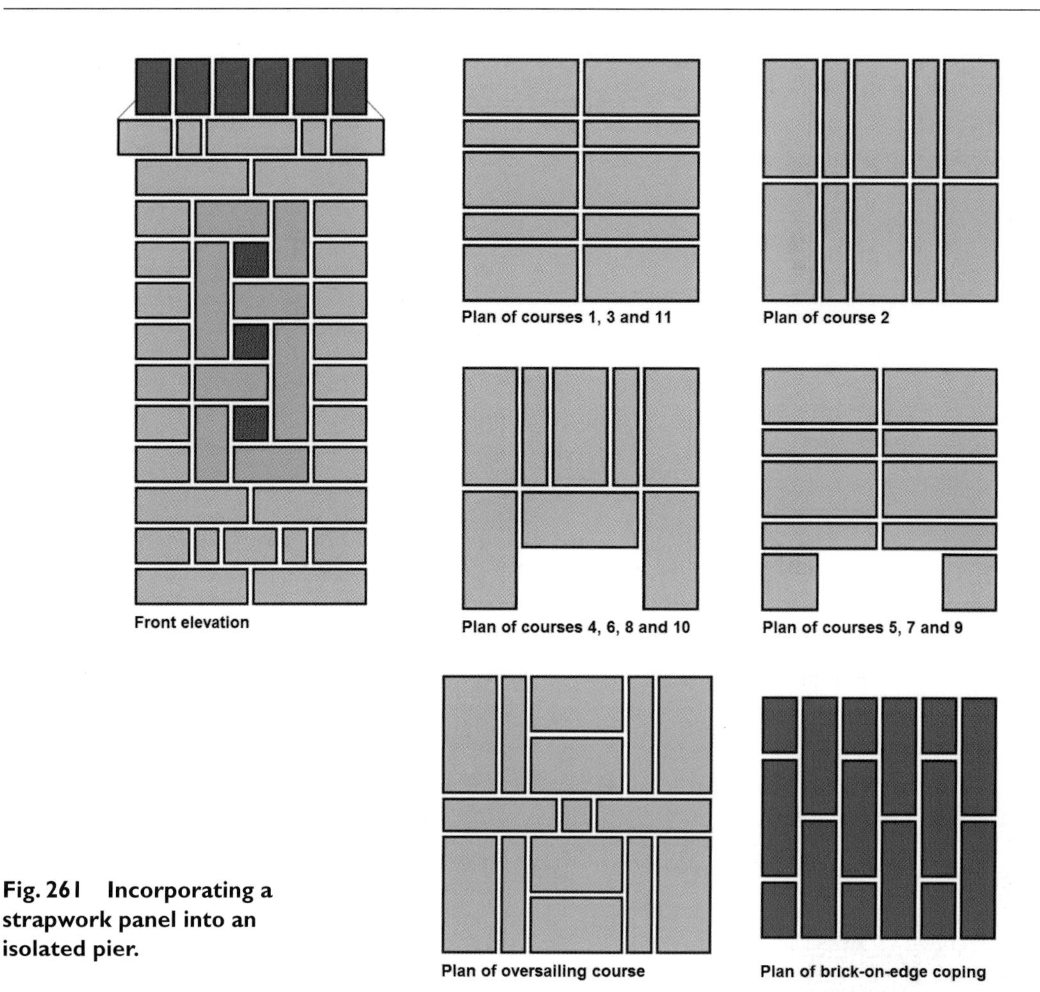

Fig. 261 Incorporating a strapwork panel into an isolated pier.

be at least one brick wide, so they can only be accommodated on piers that are a minimum of two bricks wide, in at least one direction. Panels on piers tend to be limited to one side only and no more than two opposing sides should incorporate a panel. Attempting to insert a panel on three or all four sides of a pier would compromise the bonding arrangement too much and undermine the pier's strength.

In the construction of a two-brick × two-brick (440mm × 440mm) square, isolated pier in English bond with a strapwork panel to one elevation (see Fig 261), half-bricks are used on courses 5, 7 and 9 to facilitate leaving an opening to receive the panel. If required, this could be repeated on the back of the pier without seriously compromising the bond or the strength of the finished pier.

Due to the slender width-to-height ratio of a pier, the finished panels tend to be narrow and tall, which is why they are sometimes called 'strapwork' panels.

LETTERING

Some bricklayers choose to personalize their work by introducing lettering to form words or initials (see Fig 262). The initials shown in the example involve some quite complex angled cutting in order to more faithfully represent the true shape of the letters. Such cutting is best done with a machine.

Simpler, square cuts may be used to produce more stylized, somewhat 'blocky' lettering, which, overall, is more practically achievable where access to cutting machinery might be limited. This type of lettering may also be emphasized further by using projecting bricks (*see* Fig 263).

RIGHT: **Fig. 262 Isolated pier incorporating decorative lettering.**

BELOW: **Fig. 263 Simple lettering.**

CHAPTER 14

Defects and Maintenance

BRICKWORK

Brickwork is an extremely durable building material but it will not necessarily last for ever and defects will eventually begin to manifest themselves. They may be a result of poor specification or detailing at the time of original construction or simply due to the ravages of weather over extended periods of time. Brickwork will, therefore, require maintenance from time to time.

EFFLORESCENCE

The term 'efflorescence' is used to describe the depositing of soluble salts (in other words, those that dissolve in water) on the surface of finished brickwork. Salts of magnesium, calcium, potassium or sodium may have been present in the clay used to make the bricks or in the sand used to make the mortar.

When the brickwork becomes wet, the salts dissolve into a solution. When the brickwork dries out and the moisture evaporates out of the wall, the salts revert to solid form but are left on the outside face of the material, usually in the form of a white powder.

Fig. 264 Effloresence.

Efflorescence can be seen on brickwork of any age but is most commonly found on new work that is drying out after the construction process. Efflorescence can also originate from the ground, with soluble salt solutions in the earth being absorbed by the brickwork. This explains why efflorescence commonly appears below horizontal DPC level, particularly where bricks with a high rate of water absorption have been used instead of engineering bricks.

The only real solution to efflorescence is periodically to brush off the salt deposits from dry brickwork as they come to the surface until, over time, all the salts have been released. It is a mistake to try to wash off the efflorescence as this will merely re-dissolve the salts and wash them back into the brickwork, ready to emerge again when the brickwork dries out.

New bricks are tested by manufacturers for the extent to which they are likely to effloresce and bricks are graded on the following scales in terms of the bricks' exposed surface area:

- Nil: no perceptible efflorescence.
- Slight: no more than 10 per cent displays a thin covering.
- Moderate: thin covering affecting between 10 and 50 per cent.
- Heavy: heavy deposits affecting more than 50 per cent but with no flaking.
- Serious: heavy deposits displaying surface powdering/flaking, which increases during wet weather.

It is rare to encounter bricks that are liable to effloresce any more than slightly.

Crypto-efflorescence is a second, more serious form of the effect, which can cause physical damage to brickwork. It occurs where salt crystals form just beneath the brick surface. These crystals constantly expand or contract according to whether the brickwork is wet or dry. In time, this movement can cause flaking and spalling of the brick surface. This is common in bricks that are already weak through under-firing at the point of manufacture. Crypto-efflorescence and frost damage sometime get confused with each other, as the physical damage to the brickwork resulting from both is much the same.

LIME STAINING

Lime staining is often mistaken for efflorescence, but it usually emanates from mortar joints rather than from the bricks themselves and does not disappear when wet. Like efflorescence, it needs saturation to dissolve and transport the material, in this case free lime (calcium hydroxide) present in mortar. As the water leaches and evaporates out, it is deposited on the surface, leaving a trailing effect (see Fig 265). The similarity with efflorescence ends at this point, however, since the calcium hydroxide reacts with the carbon dioxide present in the air to form calcium carbonate, which is the chemical basis for limestone and does not dissolve in water.

Lime staining is quite difficult to remove but fresh stains that have not started to carbonate can be scrubbed with a stiff brush and water, ensuring that care is taken not to damage the face of the brickwork. For older stains that have been exposed to the air for a longer time, and have begun to carbonate, thus forming limestone, acid treatment will be necessary. Pre-dampen, but do not saturate the wall with water, so as to reduce its suction. This is so that the acid is not drawn into the brickwork and stays near the surface where it can do its work. Carefully apply a proprietary, acid-based brick-cleaning solution with a paint brush to dissolve the lime, and then lightly scrub with a stiff brush and water. Such cleaning chemicals are commonly referred to as 'brick acids' and must always be used entirely in accordance with the manufacturer's instructions. This includes taking all the specified safety precautions. It is also a good idea, before treating the wall, to test the chemical on an inconspicuous area of brickwork to ensure that it will have no adverse effects on the walling.

Fig. 265 Lime staining.

If lime staining has not been completely removed after three treatments, it is unlikely that there will be any significant improvement with further applications. Under these circumstances, advice should be sought from a specialist contractor.

In order to reduce the possibility of lime staining (and efflorescence), the following points should be considered both before and during the construction process:

- Keep bricks dry when being stored and protected from rain during the construction process.
- Ensure that DPCs are correctly installed.
- Protect new and partially constructed brickwork from the rain.
- Ensure that mortar joints are full with no internal voids in which water can collect.
- Ensure that the most appropriate brick type is specified for situations where the bricks will be repeatedly wet or saturated. For example, use engineering bricks between finished ground level and DPC level or for the copings on top of boundary walls.
- Take care when detailing structures that are in contact with concrete or stone, which could allow free lime to migrate into the adjacent brickwork.

MORTAR BEES

Mortar bees (or masonry bees) are so called because they sometimes burrow into the soft mortar joints of old brick walls. There are a number of different species of bee that do this, but the most common has the scientific name of *Osmia Rufa*. In appearance it is rather like a small version of the ordinary honey bee but a little more hairy. Being small, they are often mistaken for wasps, but they are soft brown and yellow coloration rather than bright yellow and black.

The natural habitat of the mortar bee is earth banks and soft exposed rocks into which the female bee burrows a small chamber (approximately 20mm deep), to lay her eggs. Some species (often referred to as miner bees) can excavate much deeper to form more complex tunnels and galleries. Soft mortar in old brickwork makes an ideal, alternative nesting site. Contrary to popular belief, mortar bees do not eat the mortar, but merely excavate into it. Tell-tale signs of activity are small piles of mortar dust at the base of the wall. Mortar bees will also lay eggs in old drilled holes in brickwork, gaps around window or door frames, holes in airbricks and even redundant key holes. When nesting in brickwork, mortar bees are most likely to make use of south-facing walls that receive sun for most of the day.

The adult bees live for a short period of time (approximately April to July), with only one brood raised each year and nesting taking place in early spring. The bees lay their eggs in their chambers, which they also stock with pollen and nectar, following which the chamber is sealed. The eggs hatch out as larvae that feed on the pollen and nectar left in the chamber. The larvae then pupate and subsequently hatch out as bees, which emerge from the chamber.

Mortar bees are solitary creatures that, unlike honey bees or wasps, do not form social colonies. They do, however, exploit the same suitable nesting sites, giving a false impression of a larger colony. They are not aggressive, and pay little or no attention to people. They have a sting that is unable to pierce the human skin and is therefore harmless and, like all bees, are beneficial as pollinators of plants.

In terms of preventive measures, the spraying of insecticide is generally ineffective as the bees are found only on sunny elevations and insecticides break down with UV light. Moreover, the eggs are sealed inside the chambers so are protected from the insecticide. The only effective way of preventing these bees is to re-point areas of soft and perished mortar, as the bees are able to burrow only into comparatively weak materials. This work is best done in late summer or early autumn, once the bees have ceased their activities but before winter sets in, which can cause more damage to the masonry.

REPLACING PERISHED BRICKS

One of the most common problems associated with bricks and brickwork is that of 'spalling' or 'spelching', long after construction has been completed.

Fig. 266 A mortar bee.

Fig. 267 Perished and spalled brickwork.

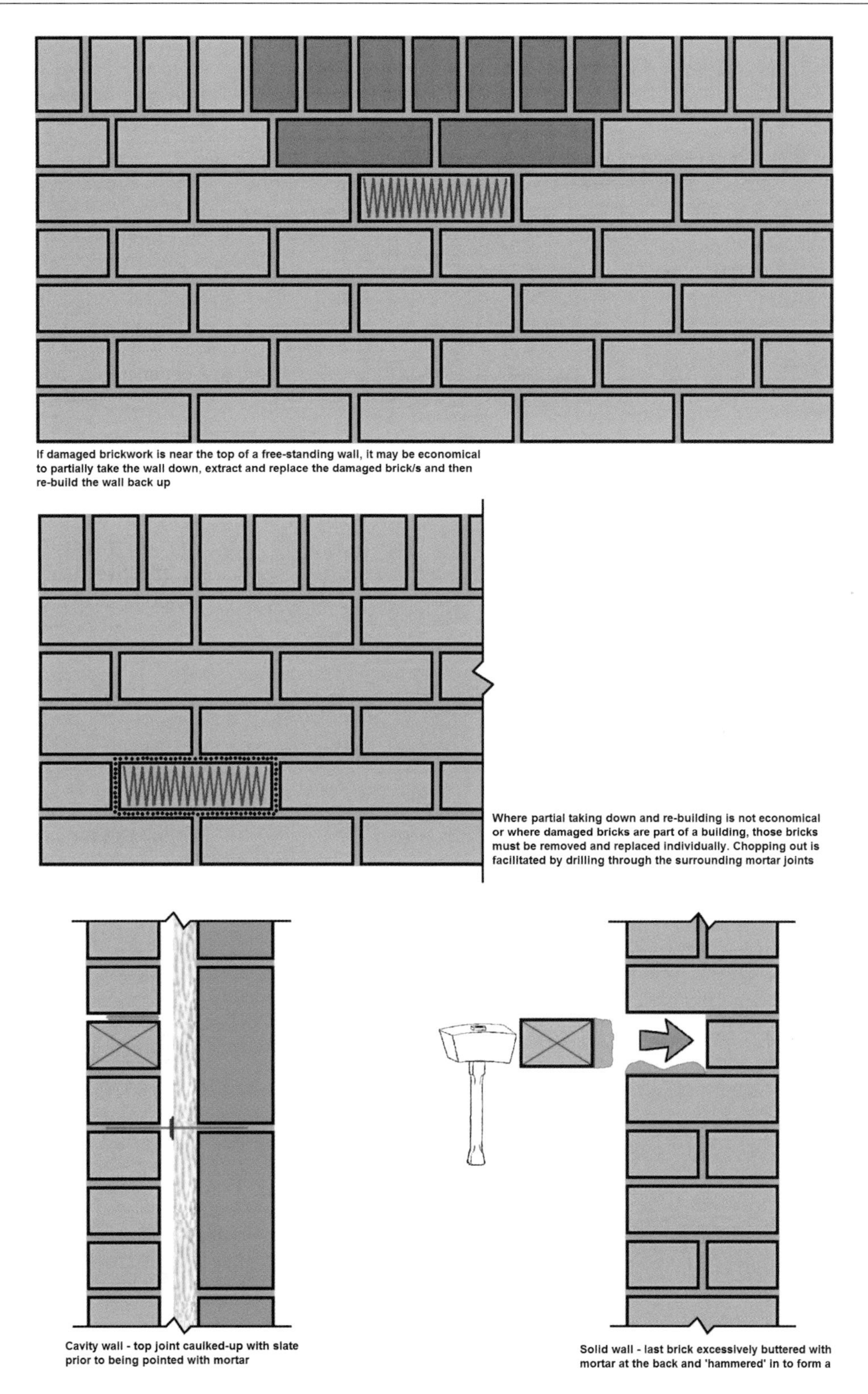

Fig. 268 Key aspects of replacing damaged and perished bricks.

This may be caused by frost damage – water in the brick freezes and then expands, and this can force the face off the brick (*see* Fig 267). A less common cause is crypto-efflorescence, but the net effect is the same. This kind of damage is most prevalent in brickwork below DPC level, on very exposed walls or on free-standing walls where rainwater is not thrown clear by the coping and simply runs down the wall face.

As part of the maintenance of external walls, it may be necessary to cut out and replace individual bricks or small areas of brickwork that have suffered spalling. The removal of damaged bricks is usually achieved by chopping out the mortar joints that surround them using a lump hammer and a sharp plugging chisel. This process is made much quicker and easier, particularly on cavity walls, by drilling holes through the mortar joints at frequent intervals with an 8mm masonry drill bit. Damage to surrounding bricks should be avoided as far as possible, which is why it is not a good idea to use anything bigger than an 8mm bit or a heavy-handed tool such as a cold chisel.

Replacement bricks should, obviously, match the existing brickwork and the new mortar should be of the same mix as the original. Any significant difference in the mix proportions between new and old mortar could result in differential movement and cracking. From an aesthetic point of view it should be noted that an invisible repair will be impossible due to introducing new materials into a wall that has weathered and, therefore, changed colour somewhat over time.

Having removed the old bricks and mortar, ensure that all the dust and debris has been removed and dampen the area with water to ensure good adhesion for the new mortar.

When replacing an individual brick, or indeed the last brick in a group, it can be difficult to achieve a tight fit and a compact and full mortar joint on top of that brick. This is most significant when repairing a cavity wall since there is nothing but air behind the brick against which to 'pin' the new mortar for the top joint, and pointing mortar simply falls into the cavity. To overcome the problem, it is necessary, having positioned the last brick, to 'caulk up' the top joint with pieces of slate. This ensures a strong, tight fit and allows the joint to be pointed up easily. At all times, care must be taken to ensure that no debris falls into the cavity.

The situation is much easier when replacing bricks at the front of a solid wall that is one brick thick or more. In order to achieve a tight joint at the top of the new or last brick, it is necessary to excessively butter the back of the brick with mortar and to force the brick into the opening. This is best achieved with a wooden or rubber mallet that will not damage the face of the brick. As the brick is forced in, the mortar at the back gets squeezed forward over the top of the brick – the mortar needs to have a high degree of workability for this to be effective. A tight joint should then be achievable with some final pointing up, without the need for any additional caulking up with slate.

For free-standing walls, there is a second option for replacing damaged brickwork. It may be economically viable partially to take down the wall to the point where the defective brick/s occur, utilizing the principle of racking back, to renew the defective brick/s and then to rebuild the wall back up. A value judgement would need to be made on a case-by-case basis as to whether such a course of action is practical, since the lower down the wall the defective bricks are located, the more significant the amount of brickwork that needs to be taken down. The time and cost of rebuilding must also be taken into consideration. If any original bricks are to be reused, they must be cleaned of old mortar ('dressed'), which will demand more time, and will further impact on any decision relating to viability.

REPAIRING PERISHED MORTAR JOINTS

Over long periods of time and prolonged exposure to the weather, mortar joints begin to deteriorate. This is more prevalent on old walls or buildings, which are likely to have been constructed with softer lime mortar and/or weaker mortar mixes. With successive winters, the mortar begins to succumb to the actions of frost, which causes the front of the joints to crumble and fall away. If left unattended, the mortar joints can deteriorate to the point where rain penetration through the wall becomes an issue, as does the risk of structural failure.

Fig. 269 Perished mortar joints.

Raking Out

The only solution is to remove the degraded mortar to a depth of at least 15mm and re-point the brickwork with new mortar. Removal of the old mortar between joints must be done carefully in order to avoid damage to the arrises of the bricks. A hammer and plugging chisel are preferred to an angle grinder, particularly on very old buildings, where the bricks are comparatively soft, and under-fired in many cases. Very soft mortar can probably be removed with an old flat-head screwdriver. Where joints are particularly narrow, as is often the case on very old brickwork, it may be necessary to remove the mortar with the blade of a masonry saw. While a small-bladed angle grinder undoubtedly provides the fastest method of mortar removal, it is difficult to keep the grinder aligned with the joints and the chance of damaging the brickwork is very high, so raking out by hand should be favoured wherever possible.

When raking out, work downwards from the top of the wall and work on approximately three courses of bricks at a time. Always remove the mortar from cross-joints first, followed by the bed joints. If the bed joints are removed first, there is a risk of chipping into the bricks above or below when one comes to clear the vertical cross-joints.

The long-term success of raking out and re-pointing depends enormously on achieving a raked-out depth of at least 15mm and making sure that all the mortar has been removed from the edges of the bricks inside the joints. Failure to do this will result in poor adhesion and the new mortar being susceptible to the action of frost; it could simply 'pop out' after only one winter. Undertaking re-pointing but failing to rake out properly, or to a sufficient depth is, therefore, a waste of time and effort! Where mortar is found to be loose and degraded beyond a depth of 15mm, it should be cleaned back until reasonably firm mortar is located.

Mortar for Re-Pointing

The mortar used for re-pointing should generally match the strength of the bricks and be a mix of cement, lime and sand. Suggested mix proportions for most applications would be 1:2:8. Walling that is more exposed to the weather would probably require a stronger mix, such as 1:1:6. Avoid using a harsh sand/cement mix that omits lime, as the mortar will set too quickly and will form a weak bond with the bricks that is easily loosened by frost. As with all mortars, the mix proportions must be accurately and consistently gauged to ensure uniformity of colour and strength.

Colour of mortar can be an important issue, particularly if only a portion of a wall is being

Fig. 270 Hand-hawk.

re-pointed, and the new mortar needs to match the existing, adjacent mortar. This is sometimes referred to a 'patch pointing'. Choice of sand has a direct influence on colour, with red sand producing a brown mortar and yellow sand producing a grey mortar. However, it may be necessary to introduce colouring additives in order to get the match as close as possible. This kind of detailed work is likely to be within the context of a historic and/or listed building and may even involve building a small temporary wall with raked-out joints (a sample panel), in order to test the colour match of the new mortar. Where listed buildings or buildings in conservation areas are concerned, it is quite common for the relevant Local Authority to specify the building of a sample panel in order to agree the mortar mix and colour to be used.

From a practical point of view, the water content of the mortar needs to be less for re-pointing than for bricklaying, simply because such a level of workability is not required or desired. The mortar for re-pointing must be firm enough to be cut into strips that will adhere to a pointing trowel and maintain their shape for long enough to be inserted into the joint, and then to receive a joint finish within a short period of time. If it is too wet, it will be difficult to push the mortar into the joints without staining the brickwork. The mortar will then have to be left for a period of time to 'go off' before a joint finish can be applied. As a test of consistency, the mortar should stand up on the pointing trowel without sagging.

Re-pointing is a slow, time-consuming process so only small amounts of mortar should be mixed at a time, otherwise it will start to go off before it is used. Batches of approximately half a bucketful are usually sufficient.

Re-Pointing Method

Re-pointing work should be carried out only in favourable weather conditions – never in wet or frosty weather or when rain or a frost is expected.

After raking out the old mortar, and in readiness for re-pointing, all dust and debris must be removed from the joints with a brush. Any left behind will interfere with the ability of the new mortar to bond into the existing brickwork. To ensure good adhesion for the new mortar, dampen the raked-out joints with a wet brush or a fine water spray, making sure that both the edges of the bricks and the old mortar are damp. The wall must not be too wet or saturated as this would lead to mortar stains on the brickwork when applying the new mortar to the joints.

For small areas of brickwork, the brick trowel blade can be loaded with pointing mortar but for larger areas its is recommended to use a hand-hawk, which can hold more mortar.

Whichever tool is being used to hold the mortar, the back of a pointing trowel is used to flatten an area of the mortar with a 'patting action', to a thickness of around 10mm (the width of a mortar joint) and with a fairly straight edge at the front. Patting the mortar in this way also helps the mortar adhere to the brick trowel or hand-hawk.

The process of filling the joints with mortar has four distinct steps (*see* Fig 271): Always starts with bed joints followed by cross-joints. Work downwards from the top of the wall, concentrating on an area of approximately 1 square metre of walling at a time.

1. Tilt the hand-hawk away from the wall and use the long inside edge of the pointing trowel blade to cut away a piece of mortar around 15mm thick (in other words, the raked-out depth to be filled) from the front of the flattened mortar. As part of the same movement, move the trowel towards the wall, while at the same time pulling the hawk away from the wall. There will now be a narrow strip of mortar on the edge of the trowel. The cutting action needs to be quick in order to ensure that the strip of mortar sticks to the back edge of the pointing trowel. When re-pointing cross-joints, the strip of mortar needs to be cut off at a length of around 65mm, to fit the height of the cross-joint and to avoid staining the brick above and below the cross-joint. Mortar that is cut off for bed joints can make use of the full length of the edge of the pointing trowel blade.
2. Keeping the trowel at an angle, move the mortar towards the joint to be filled.
3. Firmly push the mortar into the raked-out joint and, using a twisting action of the wrist, pull the mortar against the most convenient arris of the brickwork to one side (for cross-joints) or to the edge above or below (for bed joints), to 'clean' the back of the pointing trowel and to release the pointing mortar from it. Pull the pointing trowel away. (The choice of arris used for this purpose is not important – it will change numerous times during the course of the work and depends greatly on the left- or right-handedness of the bricklayer and whether he or she is working above head height or at low level.

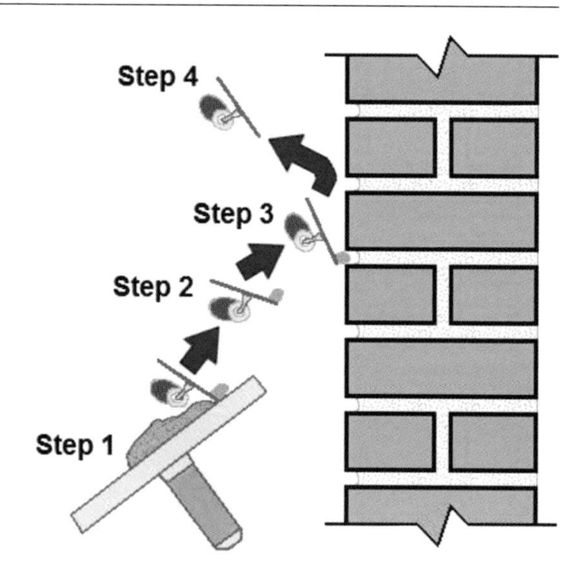

Fig. 271 Method of filling mortar joints during re-pointing.

4. To avoid staining the brickwork during re-pointing, keep the back of the pointing trowel clean by scraping it on the edge of the hand-hawk. With the empty pointing trowel, repeat the process to build up the correct amount of mortar in the joint.

When an area has been re-pointed, apply a joint finish to the new mortar; *see* Chapter 10.

EXTENDING THE LENGTH OF EXISTING WALLS

When constructing extensions or making internal alterations, there may be a need to extend the length of an existing wall. There are two recognized methods for this.

Toothing

Toothing involves cutting out every alternate brick at the stopped-end of a wall in order to build on new brickwork and continue the bonding arrangement through from the existing wall into the new section. For this reason, toothing is favoured as a method for extending walls on facing work, where the bonding arrangement will be visible and must be seen to be a continuation of the original wall.

The depth of the toothed indents cut into the existing wall will depend on the existing bonding arrangement. For quarter-bond walls the depth of the indent will be 56mm (a quarter-brick + 10mm cross-joint); for stretcher bond the depth will be 112mm (a half-brick + 10mm cross-joint).

Toothed indents should be cut into the wall working from the top down, to avoid any risk of breaking off the projecting tails of the existing bricks. The method of cutting out starts with drilling the mortar joints around the bricks to be removed and is essentially the same as for replacing perished bricks. When all the bricks and mortar have been removed the indents should be brushed out to remove any dust.

Before any new construction starts, it is vital to acquire bricks that are the same as those in the existing wall. In addition, the mortar being used for the new brickwork should match that used in the existing wall, in terms of both strength and colour. The projecting tails of the toothing on the existing wall should be checked for plumb, to establish whether any cutting is needed when joining up the new brickwork to the old – hopefully there will be no deviation. If any is present, it will be small and can be accommodated with slight adjustment to the cross-joints of the new brickwork without the need for cutting. Finally, dampen the toothing with water, to ensure good adhesion for the new mortar.

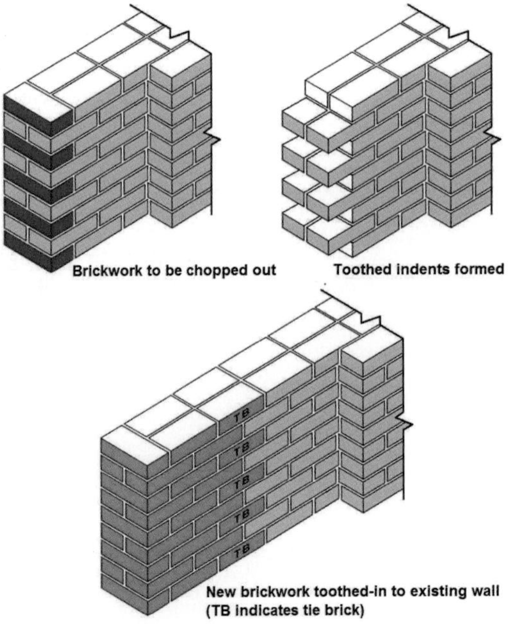

Fig. 272 Using toothing to extend the length of an existing wall.

When building away from toothings, the far end of the new wall should be set up and then the brickwork should be run-in between it and the toothings, using a string-line to ensure horizontal alignment. The string-line will most likely need to be attached to the face of the wall with a pin knocked into the mortar joints. (This will need to be made good with mortar when the work is completed.) Every tie brick (the brick inserted into the toothed indent) on alternate courses must be the first brick laid in that course and never the last. This is vital to provide the opportunity to get a full cross-joint inside the indent and to enable the bed joint above the tie brick to be caulked up solidly with mortar under the projecting tail of the existing brick above. Caulking up solidly at this point is best done using a semi-stiff mortar that can be compacted into the joint without squeezing out. Failure to achieve solid joints where new brickwork joins to old will result in a weak point within the wall.

Block Bonding

Block bonding is similar in principle and practical methods to toothing but involves cutting out indentations into the end of the wall that are either three (minimum) or five (maximum) courses high. Using this method ensures that each indent achieves an optimum number of tie bricks. Block bonding is a little quicker and easier than toothing and tends to be used to extend unseen work, where adequate strength is still needed but the aesthetics of accurate or normal bonding are less important.

The depth of the indents cut into the existing wall will, again, depend on the existing bonding arrangement. For quarter-bond walls the depth of the indent will be 56mm and for stretcher bond, 112mm. When bonding in the new brickwork at the first course of each indent it is considered good practice to insert mesh reinforcement or wall ties, for additional strength at the junction between the new and old brickwork.

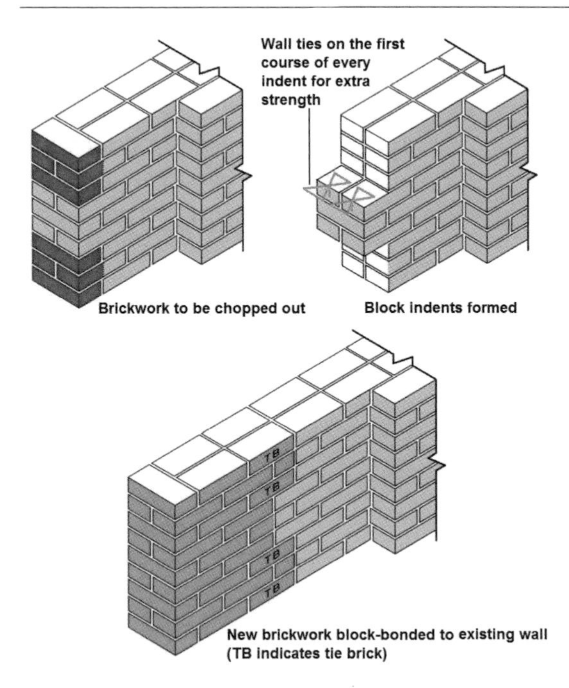

Fig. 273 Block bonding to extend the length of an existing wall, making use of indents that are half a brick deep and three courses high, incorporating two tie bricks per indent.

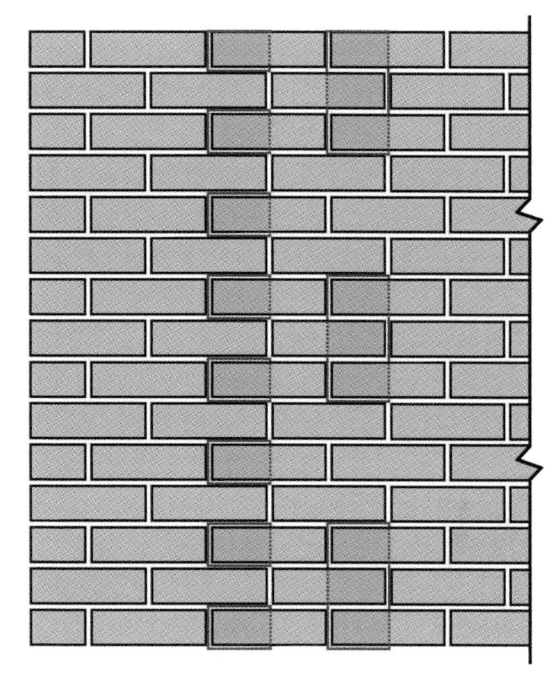

Fig. 274 Marking out the face of an existing wall for toothing in or block bonding a new junction wall.

Tying in New Junction Walls to Existing Walls

Again, when constructing extensions or making internal alterations, there is often a need to tie in a new wall at a 90-degree angle to an existing wall, in the form of a junction wall. Toothing-in and block bonding (*see above*) can be used for this purpose – the only essential difference is that the indents have to be cut into the face of the existing wall in order to tie in the new junction wall.

The first task is to mark out accurately on the existing wall the positions of the toothings or block indents to be cut out. When marking out, allow for an indent that is 25mm wider than the brick or block to be let in and take care to ensure that the indents are all vertically aligned. Having marked out the indents, the cutting process will be greatly aided by drilling and removing as much as possible of the mortar joints adjacent to the brickwork being removed (indicated by the solid red marking-out line). Where broken marking-out lines cross through bricks, these are the points at which bricks must be firmly marked with a hammer and sharp bolster chisel. This process will ensure a neat edge to the indent when chopping out is complete, which is most important when the finished work will be visible, such as when cutting toothings on facing work.

When chopping out an indent, start from the middle using a lump hammer and sharp cold chisel and work outwards, taking care not to damage the adjacent bricks when the outside edges of the

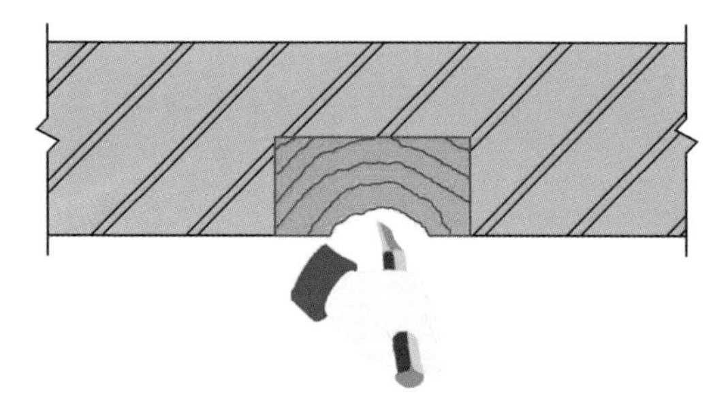

Fig. 275 Method of chopping out indents for toothing or block bonding.

Personal Protective Equipment

When using any drilling and cutting tools in relation to brickwork and masonry, it is vital to wear appropriate PPE, in the form of goggles for eye protection and gloves for hand protection. When chopping out brickwork, steel toe-capped boots should be worn as protection from falling masonry. Similarly, a safety helmet should be worn when chopping out masonry above head height.

indent are reached. Working from the outside in to the middle will result in damage to the bricks that surround the indent. When a clean edge has been achieved in the middle of the indent, it is then possible to continue chopping out with a narrow, sharp bolster instead of the cold chisel.

'Wall-Starter' Systems

Tying in a new junction wall, particularly if it is for a large extension, can be very labour-intensive and time-consuming. A quick, simple and cost-effective alternative is the use of a 'wall-starter' system. There are various versions on the market but the basic principle for all of them is much the same. Wall starters comprise two basic components: a metal channel or profile (usually stainless steel) and a set of metal ties. The profile is rawl-bolted vertically to the existing wall where the junction wall is to be joined on. Profiles come in individual lengths, from 900mm to 2m, depending on the manufacturer, and can be cut as required and/or clipped together to tie in junction walls of up to three storeys in height. Usually, one vertical profile is required for every half-brick thickness of the junction wall. The corresponding metal wall ties clip into the channel and are bedded at regular intervals into the bed joints of the junction wall as it is constructed.

As well as being quick, easy and cheaper, wall starters make it easy to tie in new brickwork of 75mm gauge to old existing walls of a different gauge that make use of bigger bricks and thinner bed joints. The only real downside is the fact that wall starters create a continuous vertical 10mm joint throughout the height of the junction and some find this unattractive. Others, however, can live with a straight joint, and are happy to make the compromise in order to save the extra time, effort and cost of cutting indents in the face of the existing wall. The straight vertical joint is usually either pointed and jointed with mortar or sealed with mastic.

Fig. 276 Staifix universal wall-starter system, a typical example of such a system, shown during construction with a new cavity junction wall being joined to an existing wall.
Ancon Building Products

Useful Addresses & Websites

Aggregate Industries
Bardon House
Bardon Road
Coalville
Leicestershire
LE67 1TD
www.aggregate.com

Ancon Building Products
Ancon Building Products
President Way
President Park
Sheffield
S4 7UR
www.ancon.co.uk

Brick Development Association
The Building Centre
26 Store Street
London
WC1E 7BT
www.brick.org.uk

Building Regulations
Department for Communities and Local Government
Eland House
Bressenden Place
London
www.communities.gov.uk/planningandbuilding/buildingregulations

Building Research Establishment
Bucknalls Lane
Watford
WD25 9XX
www.bre.co.uk

Chartered Institute of Building
Englemere
Kings Ride
Ascot
Berkshire
SL5 7TB
www.ciob.org.uk

Concrete Block Association
60 Charles Street
Leicester
LE1 1FB
www.cba-blocks.org.uk

Concrete Centre
Riverside House
4 Meadows Business Park
Station Approach
Blackwater
Camberley
GU17 9AB
www.concretecentre.com

Hanson Brick
Hanson House
14 Castle Hill
Maidenhead
Berkshire SL6 4JJ
www.heidelbergcement.com/uk/en/hanson/home

Mortar Industry Association
Gillingham House
38-44 Gillingham Street
London
SW1V 1HU
www.mortar.org.uk

Royal Institute of British Architects
66 Portland Place
London
W1B 1AD
www.architecture.com

Spear & Jackson
Atlas Way
Atlas North
Sheffield
S4 7QQ
www.spear-and-jackson.com

Stabila
Stirling Road
Shirley
Solihull
West Midlands
B90 4LZ
www.brianhyde.co.uk/stabila

Travis Perkins
Lodge Way House
Lodge Way
Harlestone Road
Northampton
NN5 7UG
www.travisperkins.co.uk

Index